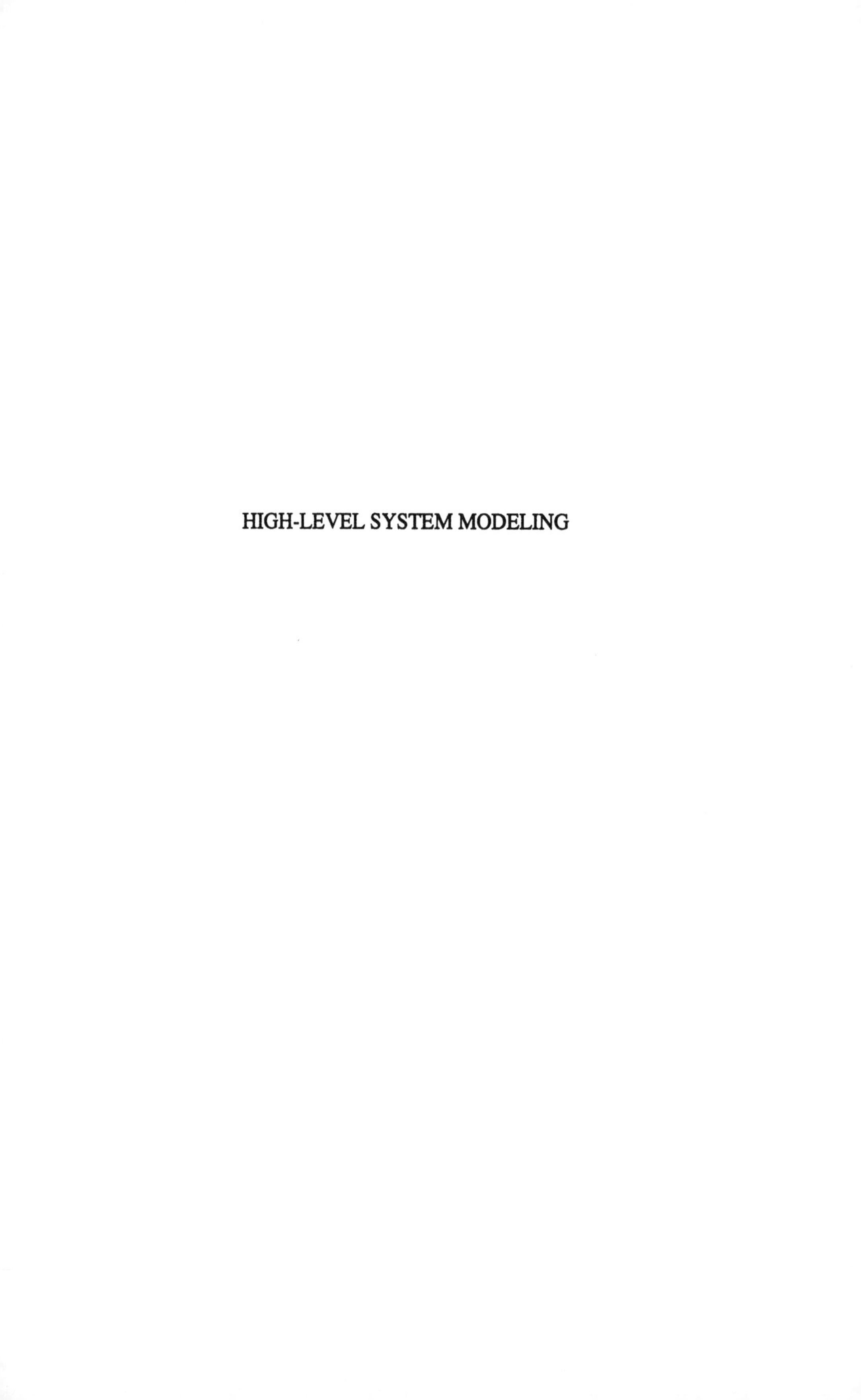

HIGH-LEVEL SYSTEM MODELING

HIGH-LEVEL SYSTEM MODELING

Specification Languages

Edited by

Jean-Michel Bergé
CNET, France

Oz Levia
Synopsys Inc., U.S.A.

and

Jacques Rouillard
ESIM, France

KLUWER ACADEMIC PUBLISHERS
BOSTON / DORDRECHT / LONDON

A C.I.P. Catalogue record for this book is available from the Library of Congress

ISBN 978-0-7923-9632-1

Published by Kluwer Academic Publishers,
P.O. Box 17, 3300 AA Dordrecht, The Netherlands.

Kluwer Academic Publishers incorporates
the publishing programmes of
D. Reidel, Martinus Nijhoff, Dr W. Junk and MTP Press.

Sold and distributed in the U.S.A. and Canada
by Kluwer Academic Publishers,
101 Philip Drive, Norwell, MA 02061, U.S.A.

In all other countries, sold and distributed
by Kluwer Academic Publishers Group,
P.O. Box 322, 3300 AH Dordrecht, The Netherlands.

Printed on acid-free paper

SERIES PRESENTATION

Current Issues in Electronic Modeling is a series of volumes publishing high quality, peer-reviewed papers dealing with modeling issues in the electronic domain. The objective is to establish a unique communication channel between academia and industry which will serve the growing needs in the field of modeling.

PUBLISHED VOLUMES:

• Volume 1: Model Generation in Electronic Modeling

***Contents**: **1.** A Flexible Generator of Component Models. **2.** What Makes an ASIC Library Sign-Off? **3.** A Case History in Building Vital-Compliant Models. **4.** Modeling Multiple Driver Net Delay in Simulation. **5.** DELPHI: the Development of Librairies of Physical Models of Electronic Components for Integrated Design Environment. **6.** VHDL Floating Point Operations. **7.** Symbolic Model Checking with Past and Future Temporal Modalities: Fundamentals and Algorithms. **8.** KRYPTON: Portable, Non-Reversible Encryption for VHDL. Index.*

168 pp. ISBN 0-7923-9568-9

• Volume 2: Modeling in Analog Design

***Contents**: **1.** VHDL-A Design Objectives and Rationale. **2.** Modeling in VHDL-A: Devices, Networks and Systems. **3.** Analog Modeling using MHDL. **4.** Modeling and Simulation of Electrical and Thermal Interaction. **5.** Modeling of Power MOSFET. Index.*

176 pp. ISBN 0-7923-9569-7

We hope you will enjoy reading this series. We welcome your suggestions and look forward to having you as a contributor.

The Series Editors

Jean-Michel Bergé, France Telecom-CNET
Email: berge@cns.cnet.fr
Oz Levia, Synopsys Inc.
Email: ozl@aol.com
Jacques Rouillard, ESIM
Email: rouillard@acm.org

EDITORS

VOLUME PRESENTATION

VOLUME 3 HIGH-LEVEL SYSTEM MODELING: Specification Languages

The process of modeling hardware involves a certain duality: a model may specify, and represent the desires and constraints of the designer, or a model may imitate something that already exists, and can end in simulation or documentation.

Surprisingly enough, one of the main qualities of a specification formalism is its ability to ignore issues that do not belong to this level. Such formalisms are obviously intended for the first stages of a design, but can also be used in the process of redesign. Having a proper level of description thus avoids two symmetric problems:

- Overspecification, which would introduce into new instances of the hardware constraints that were only meaningful to the previous ones.
- Underspecification, which would lead to unnecessary work and sometimes to starting again from scratch.

This third issue of the Current Issues in Electronic Modeling (CIEM) is dedicated to specification formalisms. The first paper explores various methods within environments that react to the model, in the context of a unified framework.

Object-oriented techniques are well-known in the world of software with the aim of allowing reuse, easier specification and packaging of the complexity; the second paper details various object-oriented methodologies.

Since the above domain still remains a topic of research, there are a number of candidate formalisms. To define their semantics, VHDL is the language of choice, either as a subset or target language. Papers 3 to 6 describe the state of the art formalisms - VSPEC which augments VHDL or ESTELLE, SDL and LOTOS - with methods that map their semantics to simulatable or synthesisable VHDL. Finally, the seventh paper is a case study used as a paradigm.

This issue, devoted to formalisms and languages, introduces volume 4 of the CIEM that presents papers focused on their use. There is little doubt that these methodologies are destined to become standard in the near future, and I am sure that you will find this issue as interesting to read as we did to edit.

Jacques Rouillard, ESIM
Co-editor of the series

CONTENTS

CONTRIBUTORS

Perry Alexander
Department of Electrical and Computer Engineering, University of Cincinnati, Cincinnati, OH 45221-0030, USA

Phillip Baraona
Department of Electrical and Computer Engineering, University of Cincinnati, Cincinnati, OH 45221-0030, USA

I.S. Bonatti
Universidade Estadual de Campinas, Faculdade de Engenharia Elétrica, Departamento de Telemática, Campinas, Brazil.

Stanislaw Budkowski
Institut National des Télécommunications, 9 rue Charles Fourier, Evry, France

James P. Cohoon
Department of Computer Science, University of Virginia, Charlottesville, VA 22903, USA

Matt Fairtlough
Formal Methods and Software Engineering Research Group, Department of Computer Science, University of Sheffield, U.K

R.J.O. Figueiredo
Universidade Estadual de Campinas, Faculdade de Engenharia Elétrica, Departamento de Telemática, Campinas, Brazil.

Mike Holcombe
Formal Methods and Software Engineering Research Group, Department of Computer Science, University of Sheffield, U.K

Florentin Ipate
Formal Methods and Software Engineering Research Group, Department of Computer Science, University of Sheffield, U.K

Camilla Jordan
Formal Methods and Software Engineering Research Group, Department of Computer Science, University of Sheffield, U.K

Carlos Delgado Kloos
Depto. Ingeniería de Sistemas Telemáticos, ETSI Telecomunicación, Universidad Politécnica de Madrid, E–28040 Madrid, Spain

.../...

Gilbert Laycock
Department of Mathematics and Computer Science,
University of Leicester, U.K.

Andrés Marín López
Depto. Ingeniería de Sistemas Telemáticos, ETSI Telecomunicación,
Universidad Politécnica de Madrid,
E–28040 Madrid, Spain

Tomás de Miguel Moro
Depto. Ingeniería de Sistemas Telemáticos, ETSI Telecomunicación,
Universidad Politécnica de Madrid,
E–28040 Madrid, Spain

Wolfgang Nebel
Department of Computer Science,
Carl von Ossietzky University,
26111 Oldenburg, Germany

John Penix
Department of Electrical and Computer Engineering, University of Cincinnati,
Cincinnati, OH 45221-0030, USA

Ambar Sarkar
Department of Computer Science,
University of Virginia,
Charlottesville, VA 22903, USA

Guido Schumacher
Department of Computer Science,
Carl von Ossietzky University,
26111 Oldenburg, Germany

Tomás Robles Valladares
Depto. Ingeniería de Sistemas Telemáticos, ETSI Telecomunicación,
Universidad Politécnica de Madrid,
E–28040 Madrid, Spain

Ronald Waxman
Department of Electrical Engineering,
University of Virginia,
Charlottesville, VA 22903, USA

Jacek Wytrebowicz
Institut National des Télécommunications,
9 rue Charles Fourier, Evry, France

Duan Zhenhua
Formal Methods and Software Engineering Research Group,
Department of Computer Science,
University of Sheffield, U.K

1

SPECIFICATION-MODELING METHODOLOGIES FOR REACTIVE-SYSTEM DESIGN

Ambar Sarkar, Ronald Waxman, James P. Cohoon

University of Virginia, Charlottesville, VA 22903, USA

ABSTRACT

The goal of this paper is to investigate the state-of-the-art in specification-modeling methodologies applicable to the design of reactive systems. By combining the specification requirements of a reactive system and the desirable characteristics of a specification-modeling methodology, we develop a unified framework for evaluating any specification-modeling methodology applicable to reactive-system design. A unified framework allows the designer to look at the spectrum of choices available and quickly comprehend the suitability of a methodology for the specific application.

Using the unified framework, we study a number of representative methodologies, identifying their respective strengths and weaknesses when evaluated for the desired characteristics. The differences and relationships between the various methodologies is highlighted. We find our framework to be quite useful in evaluating each methodology. A summary of our observations is presented, together with recommendations for areas needing further research in specification modeling for reactive systems. Two such areas are improving model continuity and providing better complexity control, especially across different abstraction levels and modeling domains. We also present a description of each methodology studied in the unified framework.

1.1. INTRODUCTION

A *reactive system* is one that is in continual interaction with its environment and executes at a pace determined by that environment. Examples of reactive systems are network protocols, air-traffic control systems, industrial-process control systems etc.

Reactive systems are ubiquitous and represent an important class of systems. Due to their complex nature, such systems are extremely difficult to specify and implement. Many reactive systems are employed in highly-critical applications, making it crucial that one considers issues such as reliability and safety while designing such systems. The design of reactive systems is considered to be problematic, and poses one of the greatest challenges in the field of system design and development.

In this paper, we discuss specification-modeling methodologies for reactive systems. Specification modeling is an important stage in reactive system design where the designer specifies the desired properties of the reactive system in the form of a specification model. This specification model acts as the guidance and source for the implementation. To develop the specification model of complex systems in an organized manner, designers resort to specification modeling methodologies. In the context of reactive systems, we can call such methodologies *reactive-system specification modeling methodologies*.

Given the myriad of specification methodologies available today, each different from the other in its chosen formalism, analysis techniques, methodological approaches etc., we establish a framework that allows one to study the merits and demerits of any specification methodology for reactive systems. To be useful, this framework should consider both the specification requirements of reactive systems and the desired characteristics of specification-modeling methodologies.

To identify the specification requirements of reactive systems, we enumerate their characteristics. During the specification process, a conceptual model for each of these characteristics is developed by the designer. The combination of the developed conceptual models results in a specification model. To enumerate the desirable attributes of a specification-modeling methodology, we broadly categorize them under three categories: supporting complexity control; developing and analyzing the specification; and maintaining model continuity with respect to other modeling stages. By combining the identified requirements of reactive-system specifications and the desired characteristics of a specification methodology, we develop a unified framework for the evaluation of the applicability of any specification-modeling methodology to reactive-system design. A unified framework allows the designer to look at the spectrum of choices available and quickly comprehend the suitability of a methodology for a specific application.

Using the unified framework, we study a number of representative methodologies, identifying their respective strengths and weaknesses when evaluated for the desired characteristics. The differences and relationships between the various methodologies is highlighted. We find our framework to be quite useful in evaluating each methodology. A summary of our observations is presented, together with recommendations for areas needing further research in specification modeling for reactive systems. Two such areas are improving model continuity and providing better complexity control, especially across different abstraction levels and modeling domains. We also present a description of each methodology studied in the unified framework. The methodologies studied are applicable to the specification of reactive systems.

1.2.CHARACTERISTICS OF REACTIVE SYSTEMS

Reactive systems follow what can be called the stimulus-response paradigm of behavior: on the occurrence of stimuli from its environment, the reactive system typically responds or reacts by changing its own state and generating further stimuli. Reactive systems are typically control dominated, in the sense that control-related activities form a major part of the reactive system's behavior.

Reactive systems have typically been contrasted with transformational systems. The behavior of a transformational system can be adequately characterized by specifying the outputs of the system that result from a set of inputs to the system. In contrast, the behavior of a reactive system is characterized by the notion of *reactive behavior* [32, 48]. To describe reactive behavior, the relationship of the inputs and outputs over time should be specified. Typically, such descriptions involve complex sequence of system states, generated and perceived events, actions, conditions and event flow, often involving timing constraints.

For example, given an adder, its behavior is easily defined as producing an output which is the sum of its inputs at any given time. This behavior is unaffected by the environment in which the adder operates, and therefore the output, given a set of inputs, remains the same regardless of time. In contrast, a traffic signal continually monitors its traffic and establishes traffic flow based on its current traffic pattern. It would be difficult to describe the behavior of the traffic signal by specifying the output (traffic signals) as a result of the input (current traffic pattern), since outputs may depend on how the traffic pattern varies over time.

We now present important characteristics of reactive systems.

- ***State-transition oriented behavior:***
  Reactive systems are intrinsically state based, and transition from one state to another is based on external or internal events. The concept of states is an useful tool to model the relationship between the inputs and outputs over time.

- ***Concurrent in nature:***
  Reactive systems generally consist of concurrent behaviors that cooperate with each other to achieve the desired functionality. Concurrency is further characterized by the need to express communication, synchronization, and nondeterminism among concurrent behaviors.

- ***Timing sensitive behavior:***
  Two categories of timing characteristics can be identified: namely, functional timing and timing constraints. Functional timings represent actual time elapsed in executing a behavior. Functional timings may change with the implementation. Guessing correct functional timing associated with a specification is therefore difficult during the specification stage. Timing constraints, on the other hand, specify a range of acceptable behavior that the implementation is allowed to exhibit. The constraints are typically externally imposed on the system, and all correct implementations of the system must obey such constraints.

  A special class of reactive systems are real-time systems, which have the added attribute of temporal correctness in addition to the functional correctness of a reactive system

- ***Exception-oriented behavior:***
  Certain events may require instantaneous response from the system. This requires the system to typically terminate the current mode of operation and transition into a new mode. In some cases, such as interrupt handling, the system is required to resume in the original state at which the interrupt occurred.

- ***Environment-sensitive behavior:***
  Since the response of a reactive system depends heavily on the environment in which it operates, a reactive system can often be characterized by the environment. For example, the specification of a data-communication network designed for handling steady network-traffic can be expected to quite different than one designed for bursty traffic.

- ***Nonfunctional characteristics:***
  Reactive systems often are characterized by properties which are nonfunctional in nature, such as reliability, safety, performance, etc. These properties are often not crucial to the system's functionality, but are often considered important enough to evaluate alternative implementations.

The expression of all the characteristics of a reactive system should be directly supported by the language of specification. Since a specification methodology is typically associated with a specific set of specification languages, the effectiveness of the methodology lies in how well its specification languages support the expression of the above characteristics. In addition to the language, the methodology plays an important role in developing the representation in a methodical rather than a haphazard manner.

1.3. SPECIFICATION-MODELING METHODOLOGY REQUIREMENTS

To appreciate the requirements of a specification modeling methodology, one must understand its role in the overall design process. The design process of a reactive system can be segmented into three major phases: the specification phase, the design phase, and the implementation phase. In the specification phase, the requirements of the system under design are formulated and documented as a specification. In the design phase, the possible implementation strategies are considered and evaluated. Finally, in the implementation phase of design, the specification is realized as a product. Note that even though the three phases may be initiated in the order we mention them, these phases often overlap. Overlapping of phases implies that during the design process, one phase may not be completed before the next phase is initiated. Another point to note is that in some methodologies, it can be difficult to distinguish the three phases from one another, especially when the difference in the levels of abstraction between the specification and implementation is small. In such cases, the specification is often a reflection of the implementation, and the process of developing specification may reflect the design phase.

We are concerned with the specification phase, where the designer (henceforth known as the specifier) typically develops a model of the system called the specification model. A specification model, or simply, a *specification* is a document that prescribes the requirements, behavior, design, or other characteristics of a system or system components. By developing and analyzing the model, the specifier makes predictions and obtains a better understanding of the modeled aspects of the system.

We now precisely define a specification-modeling methodology. A *specification-modeling methodology* is a coherent set of methods and tools to develop, maintain and analyze the specification of a given system. A method, in the context of specification modeling, consists of three components. The first component is an *underlying model* which is used to conceptualize and comprehend the system requirements. The second component is a *set of languages* that provides notations to express the system requirements. Finally, the third component of a method is a *set of techniques* ranging from loosely specified guidances to detailed algorithms that is needed to develop a complete specification from preliminary concepts.

In this paper, we focus mostly on the methods. The tools are important. However, the tools are primarily concerned with providing support for the methods. Therefore, the tools can be characterized by the method component of a methodology and are not considered separately.

There are three key requirements of a specification modeling methodology. First, it should be supported by languages that are appropriate for specifying requirements of the system. Second, it should provide assistance in controlling the complexity of specifying nontrivial systems. Third and finally, the methodology should also support the usefulness of the specification model across the design and implementation phases.

1.4. SPECIFICATION-MODELING METHODOLOGY FOR REACTIVE SYSTEMS

Based on the characteristics of a reactive system and the requirements of a specification-modeling methodology, we synthesize the necessary attributes of a reactive-system specification modeling methodology. These attributes are presented in Figure 1.

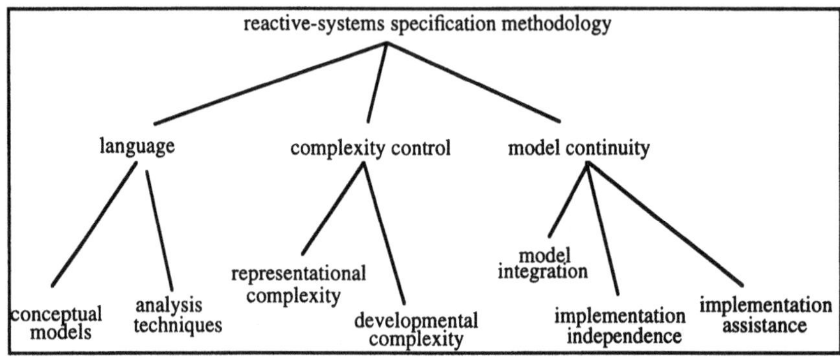

Figure 1: Attributes of a Reactive-System Specification Methodology

There are three major attributes of an reactive system specification modeling methodology: language attribute, complexity-control attribute, and model-continuity attribute.

The language attribute represents the modeling languages that supported the methodology. This attribute distinguishes reactive system specification modeling methodology from other specification methodologies, since the chosen languages should provide appropriate conceptual models and analysis techniques applicable to reactive systems. There are two dimensions of the language attribute. The conceptual-models dimension determines the available conceptual models for expressing reactive systems. The analysis-technique dimension determines the kind of support available for checking the specification consistency.

The second attribute represents the support available in the methodology for complexity control. Complexity-control is necessary for any design methodology that is applicable to design problems of nontrivial complexity. There are two dimensions along which complexity control should be supported: representational complexity and developmental complexity. Representational complexity deals with the understandability of the developed specification, whereas developmental complexity provides support for developing the specification in an organized and productive manner.

The third attribute of reactive system specification modeling methodology, support for model-continuity, distinguishes a reactive system specification modeling methodology as a specification methodology, instead of a design methodology. The specification modeling methodology should be more focussed towards developing and maintaining a specification model instead of a proposed implementation. Support of model-continuity should be considered along three dimensions: integrated modeling, implementation independence, and implementation assistance. Support along these three dimensions ensures that the usefulness of the specification model is maintained beyond the specification modeling stage of a design. Since a considerable investment of time and effort is made in developing a specification model, extending the useful life of the specification model will increase the usefulness and appeal of the associated specification methodology.

To be effective, a reactive system specification modeling methodology must strongly support all three of the attributes mentioned above. We describe each of these components in the following sections. The strengths and weaknesses of a reactive-system specification-modeling methodology can be quickly identified by evaluating the strengths of each of these components in the methodology.

1.4.1. Support for Specification Languages

The primary purpose of a specification-modeling language is to unambiguously express the desired functionality of a specified system. As discussed in Section 1.2, a reactive system possesses a number of characteristics. To express and comprehend these characteristics, a number of conceptual models are used. A specification language is defined by the conceptual models it offers the specifier to express these characteristics. In this section, we present the major conceptual models available for each reactive system characteristic. A specification language typically offers one conceptual model for a given characteristic, based on its targeted domain of application. For a good summary of available language techniques, see [9, 17, 26].

In addition to providing conceptual models, the specification language also offers support for analyzing and reasoning about the specification. The type of analysis and reasoning mechanisms that are available depend mainly on the specification language used. We think the two most important language characteristics are whether the language is based on a sound mathematical formalism and whether the specification is executable.

1.4.1.1. Available Conceptual Models

For each reactive system characteristic, we present the popular conceptual-model alternatives. When evaluating a specification-modeling methodology, one should identify if it provides conceptual models for each reactive-system characteristic. One should then check if the conceptual models that are supported by the specification-modeling methodology are appropriate given the application domain and specifier preference.

For the purpose of evaluation, we consider a methodology suitable for reactive systems specification as long as it offers at least one conceptual model for expressing the following characteristics:

- ***System views***
  There are three views that are complementary to each other: *activity*, *behavior*, and *entity* view. In the activity view, the specification represents the activities that occur in the system. Activity in a system is closely tied to the flow of data in a system. In the behavior view, the specification represents the ordering and interaction of these activities. Behavior in a system is often represented in terms of states and transitions, or the flow of control. Finally, in the entity view, the entities in the system, are identified. These entities represent the system components that are responsible for the activities and behavior in the system. Entities in a system are often represented as the data-items present in a system.

- ***Specification style***
  There are two primary styles of specification: *model*-oriented and *property*-oriented [56]. In a model-oriented specification, the system is specified in terms of a familiar structure such as state-machines, processes, or set theory. In a property oriented specification, the system is viewed as a black-box, and the properties of the system are specified in terms of the directly observable behavior at the interface of the black-box. Generally speaking, model-oriented specifications are considered easier to understand than their property-oriented counterparts. On the other hand, property-oriented specifications are considered less implementation dependent (discussed in Paragraph 1.4.3.3), since no assumption is made about the internal structure or contents of the system.

- ***Concurrency***
  Reactive systems generally consist of concurrent behaviors that cooperate with each other to achieve the desired functionality. Concurrency is further characterized by the need to express communication and synchronization among concurrent behaviors.
  Communication among concurrent behaviors
  Communication between concurrently acting portions of a system is usually conceptualized in terms of *shared-memory* or *message-passing* models. In the shared-memory model, a shared medium is used to communicate information. The communication is initiated by the sender process writing into a shared location, where it is available immediately for all receiver processes to read. In the message-passing model, a virtual channel is used, with "send" and "receive" primitives used for data transfer across that channel. Both shared-memory and message-passing models are inter-convertible: each model can be expressed in terms of the other model.
  Synchronization among concurrent behaviors
  In addition to exchanging of data, the concurrent components of a system need to be synchronized with one another. Such synchronization is needed since the concurrent components often need to coordinate their activities, and have to wait for other components to reach certain states or generate certain data or events. Synchronization may be achieved using control constructs or communication techniques. Examples of control constructs are fork-join primitives, initialization techniques, barrier synchronizations etc. Examples of using communication

techniques for synchronization are global-event broadcasting, message passing, global status detection etc.
In addition to communication and synchronization, a specification language often supports expression of nondeterminism among concurrent behaviors [21, 46]. We consider nondeterminism an attribute for complexity control rather than a reactive system characteristic, since it allows the specifier to focus on the allowable alternative behaviors without committing to any specific choice. This choice is determined at a later stage depending upon the implementation.

- ***Timing constraints***
Timing constraints are an important part of any reactive system behavior and can be specified either *directly* or *indirectly*. Timing constraints can be specified directly as inter-event delays, data rates, or execution time constraints for executing behaviors. Supporting temporal logic [46] can be seen as a direct specification technique. Indirect specification of timing constraints occurs when the actual constraint is implied through a collection of specification language constructs. This approach is followed in Statecharts [29], where timing constraints are specified by a combination of states, transitions, and time-outs.

- ***Modeling time***
Reactive system behavior is often specified in terms of their time metric (e.g. wait for 5 minutes to warm up electromechanical equipment). Consideration of time in a separately modeled entity increases the ease with which the specifier can specify such timing behavior, instead of referring to time indirectly.

- ***Exception handling***
Certain events may require instantaneous response from the system. This requires the system to typically terminate the current mode of operation and transition into a new mode. In some cases, the system is required to resume in the original state at which the interrupt occurred. Interrupt handling is one such case.
Exception handling can be provided through explicit specification language constructs. Examples are text-oriented exception handling mechanisms in Ada programming language or graphics-oriented mechanisms such as complex transitions and history operators in Statecharts.

- ***Environment characterization***
Since the reactive system is expected to be in continual interaction with its environment, it is important to be able to characterize the environment. Since the environment itself is reactive in nature, one may choose to model it using the same specification language. The operational environment of a reactive system can therefore be specified as an explicit *model* or as a set of *properties*. When specified as a separate model, the environment is seen as a separate entity that interacts with the model of the system under design. For property oriented characterization, the system's operational environment can be specified via hints about various operational conditions such as inter-arrival of events (expected frequencies, timings), expected work-loads, etc.

- ***Nonfunctional characterization***
  In addition to providing conceptual models for specifying the functionality of the system, a specification methodology reactive system should also provide support for expressing nonfunctional characteristics such as maintainability, safety, availability etc. Mechanisms to represent hints, properties, or external constraints that a system should follow should be provided.

1.4.1.2. Analysis Techniques

There are two main considerations for the support of the methodology for analyzing and comprehending the specified reactive system behavior: support for *formal analysis* and support for *model executability*.

First, we examine whether the language itself has a sound mathematical basis. Having a sound mathematical basis for the specification language enables one to automatically check for inconsistencies in the specification itself. Given the advances in formal techniques and the ever increasing number of safety critical reactive systems being designed, we consider that support for formal analysis of the specification is important. We examine if the specification language has a strong mathematical formalism as its basis. A number of formalisms are available: Petri-nets, finite state machines, state diagram, temporal logic, process algebras, abstract data types etc. A specification methodology may offer several of these formalisms as a choice for analyzing its specification models. In this paper, we observe whether the specification-modeling methodology chose a language which has a formal basis.

Second, given the typical nontrivial complexity of the systems being designed today, executability of the specification is a big help in improving the comprehensibility and robustness of the specification. Executability also offers the ability to experiment with preliminary prototypes of the system under design which is useful to validate the specification against the requirements of the system.

1.4.2. Complexity Control

One of the main requirements of a design methodology is to be able to control the complexity of the design process. Support for complexity control in a specification-modeling methodology can exist along two dimensions. The first dimension is the representational complexity, which makes the specification itself concise, understandable, and decomposable into simpler components. The second dimension is developmental complexity, which supports the development of the specification in an incremental, step-wise refined manner.

1.4.2.1. Representational Complexity

Support for this dimension is usually dependent on the specification language chosen. We list this dimension separately under complexity control, since it allows us to separate the complexity control aspect of a specification language from its support for expressing reactive system characteristics.

- ***Hierarchy***
  The notion of hierarchy is an important tool in controlling complexity. The basic idea in hierarchy is to group similar elements together and to create a new element that represents this group of similar elements. By introducing the common behavior in this way, multiple levels of abstractions can be supported.

- ***Orthogonality***
  A complex behavior can often be decomposed into a set of orthogonal behaviors. Two behaviors are orthogonal if they can be described independently of one another. For example, the user interface of a patient-monitoring system can be described independently of how the system actually implements its functionality. By supporting such decomposition of the representation, significant improvement in clarity and understandability can be attained.

- ***Representation scheme***
  The representation scheme plays an important role in the understandability of a specification. We make the distinction between *graphical* and *textual* representation schemes. By graphical scheme, we imply visual formalisms where both syntactic and semantic interpretations are assigned to graphical entities. For example, Statecharts, Petri-nets etc. are such visual formalisms. In our opinion, graphical representation schemes are preferable to a textual ones since the former allow the specifier to visualize the system behavior more effectively, especially during execution of the specification. For many textual approaches, however, tools exist today that transform the textual approach into a graphical approach.

1.4.2.2. Developmental Complexity

In addition to the representation of system behavior, a specification methodology must also support the evolution of the specification model from initial conceptualization of system requirements. We think the following dimensions should be supported.

- ***Nondeterminism***
  By incorporating nondeterminism in a controlled manner, the specification can leave details to the implementation and final stages. For example, in a typical producer-consumer type system, if requests for both element insertions and element deletions from a buffer arrive simultaneously, the specification may non-deterministically select either operation to be executed first. The commitment to an actual choice in such a scenario is deferred to later design stages. Nondeterminism thus supports a evolutionary approach to specification development. In addition to the incorporation of nondeterminism, a specification methodology should also provide mechanisms to detect and resolve nondeterminism. It is often difficult to detect nondeterminism in specifications of complex systems, given the complex interrelationships of the behaviors of the system components. If left undetected and consequently unresolved, nondeterminism is a potential source of ambiguity.

- ***Perfect-synchrony hypothesis***
  Perfect-synchrony hypothesis [5] implies that a reactive system produces it outputs synchronously with its inputs. In other words, the outputs are produced instantaneously after the inputs occur. This assumption of perfect-synchrony is borne out in many cases, especially if the inputs change at rates slower than the system can react. For example, in a clocked system, if the clock cycle is long enough, the system gets a chance to stabilize its output values before the next clock event occurs.
  Based on the assumption of perfect-synchrony hypothesis, specification languages can be divided into two types: *synchronous* and *asynchronous*. In synchronous languages, time advances only if explicitly specified. In asynchronous languages, time advances implicitly. Examples of synchronous languages are Statecharts, Esterel etc. Examples of languages based on asynchronous hypothesis are those based on concurrent programming languages such as Ada, SREM etc.
  Perfect-synchrony assumption makes the specification concise, composable with other specifications, and in general lend the specification to a number of elegant analysis techniques.
  However, the assumption of perfect synchrony may not be valid at all levels of abstractions especially if the reaction of the system is complicated. For example, if the reaction is a lengthy computation, clearly assumption of perfect synchrony will be violated, as the input may change before the computation is completed.

- ***Developmental guidance***
  Guidance for model development is helpful in identifying the next step in the process of specifying a system. One may do it *bottom up*, where the primitives are first identified and then combined. Another guidance is in the form of a *top-down* approach, where a specification is decomposing into smaller and more detailed components. Yet another approach is called the *middle-out* approach, which combines both top-down and bottom-up approaches.
  Typically, development style is a matter of choice and not strictly enforced. In some methodologies, however, this can be enforced. Enforcing the style may interfere with designer creativity.

1.4.3. Model Continuity

Significant effort is involved in the development and debugging of a model of the system under design. Once the model has been developed and analyzed, however, it is often discarded and is not revisited in the remainder of the design process. Such a limited useful life of a specification model tends to make the corresponding modeling methodology unpopular with designers. The reason for the limited usefulness of a specification model is the problem of maintaining model continuity.

Model continuity can be defined as the maintenance of relationships between models created in different model spaces such that the models can interact in a controlled manner and may be utilized concurrently throughout the design process. The problem of maintaining model continuity for a specification can be divided into following three subproblems: model integration, implementation assistance, and implementation independence. Model integration addresses the challenge of making the specification model compatible with models developed during the design and implementation stages. Implementation assistance increases the usefulness of the specification by helping during the design and implementation stages. Implementation independence increases the useful life of the specification by not committing it to a particular design/implementation choice, thus avoiding restriction of creativity during the design process.

1.4.3.1. Model Integration

Model integration can be subdivided into three attributes:

- ***Conformance***
  The conformance attribute identifies how a methodology addresses the first subproblem of model continuity, namely: checking conformance among models developed. The methodology should provide either a simulation-based support or an analysis-based support for checking conformance between the models (or both).
  We categorize conformance checking along two dimensions: *vertical* and *horizontal*. Vertical-conformance checking involves validating conformance between models representing different levels of abstraction. Horizontal-conformance checking involves validating conformance between models representing different modeling domains. To be effective, the methodology must provide support for checking conformance along both dimensions. For example, one should be able to check the conformance between an algorithmic-level model and a logic-level of a system. This is a example of vertical-conformance checking. As an example of horizontal-conformance checking, one should also be able to check the conformance between a functional level behavioral model involving register-transfers and state-sequencers and a structural model involving ALUs, MUXs, registers etc.

- ***Model interaction***
  The interaction attribute identifies how a methodology addresses the second and third subproblems of model continuity, namely: maintaining visibility of the specification model during the implementation phase and incorporating relevant details obtained from the implementation phase back into the specification model. Supporting such a high degree of interaction and information flow among these models requires integrated modeling across different levels of abstraction and modeling domains. By integrated modeling, we imply that the flow of information occurs in both directions across the model boundaries. This flow of information can occur during either integrated simulation or integrated analysis of both models. A methodology must support bidirectional information flow across model boundaries.
  Analogous to conformance checking, we categorize model interaction along two dimensions: vertical and horizontal. Vertical interaction occurs between models belonging to different levels of abstractions, whereas horizontal interaction occurs between models belonging to different domains of modeling. A methodology must provide mechanisms that support both vertical and horizontal model interactions.

- ***Complexity***

  The complexity attribute addresses the problem of controlling complexity during the development and analysis of models throughout the design stages. Given the considerable complexity of models representing nontrivial systems, support for this attribute is necessary for effective implementation of the conformance and interaction attributes.

  Complexity control is primarily achieved by supporting a hierarchy of representations. Support of hierarchy significantly reduces design time, as the designer is allowed to provide less detail in creating the original representation. For adding or synthesizing further information, he or she can then use automated or semi-automated design aids. In addition to the addition or the synthesis of details, a hierarchical approach allows the designer to quickly identify what portion of the design should be expanded upon, without necessarily expanding the rest of the system. This incremental-expansion approach is of tremendous advantage when the expanded representation is radically different from the original representation. By enabling incremental modifications, a hierarchical representation improves designer comprehension of the effect of change on the original model.

  Similar to conformance checking and model interaction, model complexity can also be divided into two dimensions: vertical and horizontal. Abstraction of a lower-level model into a higher-level is an example of managing vertical complexity. Combination of models from different modeling domains into a unified representation is an example of managing horizontal complexity. The hierarchical representation must possess capability to manage both horizontal and vertical complexity.

1.4.3.2. Implementation Assistance

The task of developing an implementation from a specification is complex. As a result, there has been a significant research in the automated synthesis of implementations from specifications [9, 14, 15, 31, 41, 55, 57]. There are two prime motivations for implementation assistance: reduction of designer effort and increase in implementation accuracy. By supporting automated/semi-automated techniques for the synthesis of a design/implementation, there is a significant reduction in required designer effort. Also, an automated technique avoids human errors that can be introduced otherwise during the manual design process. Synthesis of efficient implementations from system-level specifications is still immature. Another limitation of current synthesis techniques is that they are generally based on the structure of the specification, thus limiting design space and therefore producing less optimal solutions.

1.4.3.3. Implementation Independence

According to [56], a specification has an implementation bias if it specifies externally unobservable properties of the system it specifies. A specification is therefore considered implementation independent if it lacks implementation bias. While evaluating a specification methodology, we examine how well it supports implementation independence. There are two key advantages of an implementation independent specification. First, it allows the specifier to focus on describing the behavior of the system, rather than how it is implemented. Second, it avoids placing unnecessary restrictions on designer freedom.

1.5. A SURVEY OF METHODOLOGIES

We consider ten specification methodologies and evaluate their relative merits and demerits against the framework established in Section 1.4. Clearly, a number of other specification methodologies exist that have not been included in this report. However, we consider the presented set to be representative.

A few observations regarding these ten methodologies are in order. The opinions presented are subjective, especially when considering issues such as complexity control and model continuity. Many of the deficiencies pointed out are not necessarily fundamental to the methodologies themselves, but, in our judgement, need to be explicitly addressed by them. Each methodology therefore has the potential to evolve to the point where they support all the dimensions indicated in the framework. No relative judgement has been offered between any two methodologies.

The surveys are organized as follows. We introduce each methodology with a brief background of its development. We next mention the system views supported, followed by a brief description of the steps in the methodology. We conclude each survey with a description of a few noteworthy features that we consider are the methodology's strengths and weaknesses. These surveys are by no means exhaustive, and the reader is encouraged to follow up the references provided for further detail. The main features of the methodologies are summarized in Tables 1-4.

In each table, the rows represent the methodology and the columns represent the attributes. Each table cell presents a summary of how the corresponding specification methodology supports the attribute represented by the corresponding column. In Tables 1, 2, and 3, we describe the support of the surveyed specification methodologies for the language, complexity control, and model continuity attributes respectively. The three attributes are then summarized in Table 4.

1.5.1. Ward and Mellor's Methodology (SDRTS or RTSA)

Ward and Mellor's methodology [54], called Structured Development for Real-Time Systems (SDRTS), was developed for the specification and design of real-time applications. Since the methodology is an extension of the Structured Analysis [18] methodology for real-time systems, it is also called Real-Time Structured Analysis (RTSA). A similar methodology can be found in Hatley and Pirbhai's [31] approach.

There are two system-views adopted by RTSA: Data-Flow Diagrams (DFD) and Control-Flow Diagrams (CFD). DFD is used to model the activities in the system, whereas the CFD is used to express the sequencing of these activities. The CFD itself is defined in terms of a finite-state model such as FSM or decision table.

The methodology is based on structured analysis, one of the earlier approaches to express system requirements graphically, concisely, and minimally redundant manner. To develop a specification of the system, two models are created: the environment model and the system model. The environment model describes the operational context in which the system operates and the events to which the system reacts. The system model, which expresses the system's behavior, is then developed through successive refinements.

The language attribute is not supported well, since several aspects of reactive system behavior cannot be modeled conveniently. For example, there is no direct support for specifying timing constraints or handling exceptions elegantly. A lack of any formal method support makes the methodology less useful in specification and design of critical systems. Lack of orthogonality makes it harder to conveniently represent and understand the behavior of complicated systems. The model-continuity attribute is also not well supported.

1.5.2. Jackson System Development (JSD)

Jackson System Development [39] methodology, originally developed for program design [40], is considered suitable for the design of information systems and real-time systems [11]. In its current stage, the methodology covers specification, design and implementation stages.

The methodology is based on entity modeling. The system requirements are expressed as Jackson's diagram for entities. The diagram presents a time-ordered specification of the actions performed on or by an entity. A system specification diagram is also created, which is a network of processes that model the real world. A process communicates with others by transmitting data and state information. It is also possible to model time explicitly by introducing explicit delays.

The basic modeling philosophy is that the structure of a system to be designed can be determined from the structure and evolution of data it must manage. The methodology consists of two phases: specification and design. In the specification phase, the environment is described in terms of entities (real world objects the system needs to use) and actions (real world events that affect the entities). These actions are ordered by their expected sequence of occurrences and represented with Jackson diagrams. The actions and entities are then represented as a process network using system specification diagrams. The connection between these processes and the real world are defined. The creation of the initial process network can be seen as the end of specification phase.

During the design phase that follows the specification phase, the process network is successively elaborated by identifying further processes that are needed to execute the actions associated with the entities described in the Jackson Structure Diagram. The completed process network represents the final design which is then mapped to a set of hardware/software components.

The JSD methodology supports model continuity by carrying the specification through further well defined design steps. However, timing considerations come very late, almost after the design phase. The specification is implementation dependent, since it is closely tied to an implementation, thereby reducing designer freedom. The methodology lacks support for expressing several reactive-system characteristics such as exception handling. Neither does it support any formal analysis or well defined execution semantics. The main strength of the methodology lies in its complexity-control attribute, especially regarding its developmental guidance and representational elegance.

1.5.3. Software Requirements Engineering Methodology (SREM)

Software Requirements Engineering Methodology (SREM) [2] was developed for creation, checking and validation of specifications of real-time and distributed applications for data processing.

The SREM method is useful for specification making use of structured finite-state automata called requirement nets (r-nets). R-nets express the evolution of outputs and final state starting from inputs and current state. Both inputs and outputs are structured as sets of messages, communicated by an interface connected to the environment.

To develop a specification, the interface between the system and the environment and the data-processing requirements are specified. The initial description is produced using r-nets. Functional details, timing and performance constraints are then added. Next, validation and coherency checks are performed on the specification. A final feasibility study is conducted to guarantee that the specification will result in a feasible solution.

The behavior of a reactive system is well-represented by this methodology. The addition of performance specifications and timing constraints are also beneficial. However, hierarchical decomposition of the specification is not supported. As a result, the task of specification requires too much detail. The attribute of model-continuity is supported since both the specification and implementation can be considered in an integrated manner for testing, feasibility analysis, supporting fault-tolerance etc. Both implementation assistance and implementation independence are supported by the methodology since it allows one to study multiple implementation schemes for a given specification. However, the methodology ties the specification too closely to the possible implementation in terms of its structure. As a result, the support for model continuity can be considered to be limited, since the number of implementations considered is limited by the structure of the specification.

1.5.4. Object-Oriented Analysis (OOA)

OOA stands for Object Oriented Analysis, and is based on the object-oriented paradigm of modeling. The world is modeled in terms of classes and objects that are suitable to express the problem domain. Different schemes have been suggested for OOA [16, 48, 50], they differ mostly in terms of notations and heuristics.

OOA can be seen as an extensions of data or information modeling approaches. The latter approaches focus solely on data. In addition to modeling data, OOA also concerns data transformations.

The major steps in OOA consist of identifying objects, their attributes, and the structure of their interrelationships. The entire specification is developed as a hierarchy of modules, where each module is successively refined both horizontally and vertically into further modules. The vertical refinement adds further properties to a module, whereas the horizontal refinement identifies a set of loosely-coupled, strongly-cohesive interacting sub-modules that define the behavior of the original module. The implementation of these modules is, however, not considered at the specification stage.

The OOA main strength lies in the complexity-control attribute. To exploit the strength of OOA, however, it should be combined with languages that support expression of reactive system characteristics and have both formal and operational semantics. For application to the domain of reactive systems, we find approaches that combine languages suitable for expressing reactive systems with object-oriented design principles. Such an example can be found in [13].

1.5.5. Specification and Description Language (SDL)

Specification and Description Language is a design methodology that has been standardized by CCITT [12, 25], and is used for specifying and describing many types of systems. SDL is standardized, semiformal, and can be used both as a graphical or a textual language. SDL is primarily used for telecommunication systems [52].

SDL provides three views of a system: structural, behavioral and data. It is the behavioral view of the system that is used to specify the system's reactive nature. The system is modelled as a number of interconnected abstract machines. The machines communicate asynchronously.

The specification models are developed for the three system-views as follows. The structural model is generated hierarchically, starting from a *block* that is recursively decomposed into a number of *blocks* connected together by *channels*. At the topmost level, the *system block* has *channels* that allow interfacing with the system's environment. The behavioral model is a set of *processes* that are extensions of deterministic finite-state machines. The interaction between these *processes* is done via *signals*. These *processes* can be dynamically created and collectively represent the system behavior. Temporal ordering between the *signals* used in inter-*process* communication is specified using message-sequence charts, which are useful for debugging the specification. Finally, the data is modeled as an Abstract Data Type, where one describes the available data-operations and data-values but not how they are implemented.

SDL supports the perfect-synchrony hypothesis and has an associated formal semantics. However, it does not support all the reactive system characteristics. For example, exception handling is not directly supported since the inputs are typically consumed by a process only when the receiving process is ready to process the input. Thus, if an exception condition is communicated as an input to the process, it may not be acted upon immediately. Rather, the exception will be handled when the receiving process is ready to process the arriving exception. The complexity-control attribute is well supported. Model-continuity is not well supported in the methodology.

1.5.6. Embedded Computer Systems (ECS)

The Embedded Computer Systems (ECS) methodology [32, 41] is based on the three views of system modeling: activities, control, and implementation. Express-VHDL [37] is used as a computer-aided design tool for this methodology.

ECS supports both behavioral and functional decomposition of the system's specification. The system behavior is expressed using the visual formalism of Statecharts [30], an extension of FSM that significantly reduces the representational state-space explosion problem encountered by ordinary FSMs. This reduction is achieved due to the conceptual models of concurrency, hierarchy, and complex transitions supported by Statecharts. A history operator, useful for expressing interrupt-handling, is also provided to significantly reduce the representational complexity.

The system is functionally decomposed using activity charts, and is viewed as a collection of interconnected functions (activities) organized in a hierarchy. The activity charts visually depict the flow of information in the system, with the control of flow being represented by the associated Statecharts model.

To develop the specification, the methodology recommends a top-down and iterative analysis that gradually expresses all the requirements of the system. Conceptually, a system is decomposed into a number of subsystems, each carrying out a functionality, and a controller that coordinates the activities between these subsystems. The behavior of each system is represented by a Statecharts model. Each state in the Statecharts model can be refined further into AND and OR states. The default-entry states, needed synchronizations, associated timing constraints, etc. are specified next.

Both the language and complexity-control attributes are well supported by the ECS methodology. ECS is based on the language of Statecharts which is very suitable for expressing reactive system requirements in an implementation-independent manner. Support for the executability and some static and dynamic analysis is provided at each stage of development. However, model continuity, especially along the model-integration dimensions, is not supported.

1.5.7. Vienna Development Method (VDM)

VDM (Vienna Development Method) [41] is an abstract model-oriented formal specification and design method based on discrete mathematics [33]. The formal specification language of VDM is known as META-IV [5].

The specification is written as a specification of an abstract data type. The abstract data type is defined by a class of objects and a set of operations to act upon these objects while preserving their essential properties. A program is itself specified as an abstract data type, defined by a collection of variables and the operations allowed on these variables. The variables make the notion of state explicit, as opposed to property-based methods.

The specification development process is closely tied to the design process. The steps of the specification methodology is as follows. A formal specification is developed using the META-IV language. Once the specification is checked via formal analysis and found to be consistent, the specification is refined and further decomposed into what is called a *realization*. The realization is checked against the original specification for conformance. The specification is iteratively and step-wise refined until the realization is effectively a complete implementation.

This method supports model continuity very well. The main drawback, in our opinion, is that it is relatively difficult to understand and conceptualize a reactive system from a VDM specification. Neither is the specification executable. The structure of the specification has a large impact on the implementation, making the methodology weak on the implementation independence since the implementation is largely determined by how the specification is decomposed. It is not clear how one can incorporate independently developed implementations and check for their conformance against the specification.

1.5.8. Language Of Temporal Ordering Specification (LOTOS)

LOTOS (Language Of Temporal Ordering Specification) is an internationally standardized formal description technique, originally developed for the formal specification of OSI (Open Systems International) protocols and services.

The specification is based on two approaches: process algebras and abstract data types. The process-algebra approach is concerned with the description of process behaviors and interactions and is based on Milner's Calculus of Communicating Systems [44] and Hoare's work on Communicating Sequential Processes [34]. The abstract data type approach is based on ACT-ONE [21], which deals with the description of data structures and value expressions. The resulting specifications are unambiguous, precise, and implementation independent.

A system is specified by defining the temporal relationships among the interactions that make up its externally observable behavior. These interactions are between processes, which act as black-boxes. A black-box is an entity capable of performing both internal actions and external actions. The internal actions are invisible beyond its boundaries whereas the external actions are observable by an *observer* process. Interactions between these processes is achieved using *events*. The processes are specified using process algebra approach, which allows the description of behavior as a composition of basic behaviors using a rich set of combining rules.

Being property-oriented, LOTOS supports implementation independence, as it specifies behavior of the system that can be observed only externally. The structure of the implementation is not restricted by the specification. However, being property-oriented also makes it hard to conceptualize the internal states of reactive system. Further, it also becomes hard to generate an implementation from the specification. The concept of time is also not directly supported.

1.5.9. Electronic Systems Design Methodology (MCSE)

MCSE (Méthodologie de Conception des Systèmes Electroniques) is a methodology for the specification, design, and implementation of industrial computing systems. The methodology is characterized by its top-down approach to the design of real-time systems, and is structured into several steps from system specification to system implementation.

Three kinds of specifications are produced during the specification process. First, functional specifications include a list of system functions and a description of the behavior of the system's environment. Second, operational specifications concern the performance and other implementation details that are to be used in the system. Third and finally, technological specifications include specifications of various implementation constraints such as geographic distribution limitations, interface characteristics etc.

There are two main parts in the specification process: environment modeling and system modeling. In the environment-modeling part, the environment is first analyzed to identify the entities that are relevant to the system. Next, a model is created representing the identified entities and their interactions, thus providing a functional description of the environment. In the system-modeling part, the system under design is first delimited in terms of its inputs and outputs. Next, a functional specification of the system is developed, which describes the functions to be carried out by the system on its environment. This functional specification is developed by characterizing the system in terms of system inputs and outputs, system entities, or system activities.

The MCSE approach provides a well defined methodology for developing a specification. Another strength of this approach lies in the fact that it allows a multitude of system views and modeling approaches for system specification. However, there is a lack of support for formal techniques. While several modeling techniques and system views are supported, a coherent integration of these diverse approaches is not supported.

1.5.10. Integrated-Specification and Performance-Modeling Environment (ISPME)

The Integrated Specification and Performance Modeling Environment [49] is an evolving specification modeling methodology that supports a strong interaction between the specification phase of a design process with design and implementation phases. As a result of this interaction, there is an increased and better communication of design intent among these phases.

ISPME is based upon the language of Statecharts, similar to ECS modeling methodology. As a result, it supports the behavioral view of the system. However, ISPME also supports complementary modeling, where some aspects of the system are modeled using Statecharts, while the remaining aspects are represented as a performance model. The performance model is developed using ADEPT [2, 3], which is based on an extension of Petri-nets [47]. Since a performance-model can coexist with the Statecharts specification, ISPME supports the activity view for a system.

The methodology supports the development of a complete implementation from the specification in a incremental and iteratively refined manner. Each increment represents a proposed implementation of a component of the Statecharts model. The performance model of the proposed implementation is verified against its Statecharts counterpart. At each iteration, both the Statecharts and the ADEPT models can be refined, since it is possible that one may encounter inconsistencies between the specification and the implementation. As a result of this integrated-modeling approach, the methodology extends the specification phase to later design stages. Such extension improves communication of design intent between various phases of design.

The ISPME approach supports all three attributes of a specification modeling methodology for reactive systems. First, by adopting the Statecharts language, it is able to support both the language and the complexity attributes. Second, due to its support for integrated modeling with ADEPT, the model interaction component of the model-continuity attribute is well supported. Since the performance model developed is independent in terms of structure from the Statecharts model, implementation independence is supported. Further, implementation assistance is supported since the Statecharts model itself may be synthesized into an implementation.

1.6. SUMMARY

We summarize our conclusions drawn from studying Tables 1-4. For each table, we looked down each column representing a feature and tried to identify if the feature was well supported by the methodologies. Wherever possible, we identify correspondences between features and attempted to present general observations and recommendations.

Specification Modeling Methodologies	Support for language attribute [Section 1.4.1]							
	Available conceptual models [Section 1.4.1.1] (All methodologies surveyed here support concurrency and show limited support for nonfunctional characteristics)						Analysis techniques [Section 1.4.1.2]	
	System views	Specification style	Timing constraints	Modeling time	Exception handling	Environment characterization	Formal analysis	Model executability
SDRTS [Section 1.5.1]	activity+behavior	model	indirect	limited	limited	model	limited	limited
JSD [Section 1.5.2]	entity	model	direct	supported	limited	model	limited	limited
SREM [Section 1.5.3]	behavior	model	direct	supported	limited	property	semiformal	supported
OOA [Section 1.5.4]	entity+behavior	model	indirect	limited	limited	limited	limited	limited
SDL [Section 1.5.5]	entity+behavior	model	indirect	supported	limited	limited	supported	supported
ECS [Section 1.5.6]	activity+behavior	model	indirect	supported	supported	model	supported	supported
VDM [Section 1.5.7]	entity	model	indirect	limited	supported	limited	supported	supported
LOTOS [Section 1.5.8]	entity+behavior	property	indirect	limited	supported	property	formal	limited
MCSE [Section 1.5.9]	activity+behavior	model, property	direct	supported	limited	model	semiformal	limited
ISPME [Section 1.5.10]	activity+behavior	model	indirect	supported	supported	model	formal	supported

Table 1: Support for Language Attribute by Specification-Modeling Methodologies

Specification Modeling Methodologies	Support for complexity control [Section 1.4.2]					
	Representational complexity [Section 1.4.2.1]			Developmental complexity [Section 1.4.2.2]		
	Hierarchy	Orthogonality	Representation scheme	Nondeterminism	Perfect-synchrony Assumption	Developmental guidance
SDRTS [Section 1.5.1]	supported	limited	graphical	limited	asynchronous	top down
JSD [Section 1.5.2]	supported	supported	graphical	limited	asynchronous	top down
SREM [Section 1.5.3]	limited	limited	graphical	limited	asynchronous	bottom up
OOA [Section 1.5.4]	supported	supported	textual	limited	asynchronous	top down
SDL [Section 1.5.5]	supported	supported	graphical	supported	synchronous	top down
ECS [Section 1.5.6]	supported	supported	graphical	supported	synchronous	top down
VDM [Section 1.5.7]	supported	supported	textual	supported	synchronous	top down
LOTOS [Section 1.5.8]	supported	supported	textual	supported	synchronous	top down
MCSE [Section 1.5.9]	supported	supported	graphical	limited	synchronous	top down
ISPME [Section 1.5.10]	supported	supported	graphical	supported	synchronous + asynchronous	top down bottom up

Table 2: Support for Complexity-Control Attribute by Specification-Modeling Methodologies

Specification Modeling Methodologies	Support for model-continuity attribute [Section 1.4.3]				
	Model integration [Section 1.4.3.1]			Implementation independence [Section 1.4.3.3]	
	Conformance	Interaction	Complexity		
SDRTS [Section 1.5.1]	limited	vertical	vertical	limited	limited
JSD [Section 1.5.2]	limited	vertical	vertical	limited	supported
SREM [Section 1.5.3]	limited	vertical	vertical	supported	limited
OOA [Section 1.5.4]	limited	limited	limited	limited	supported
SDL [Section 1.5.5]	limited	limited	limited	supported	supported
ECS [Section 1.5.6]	limited	limited	limited	supported	supported
VDM [Section 1.5.7]	supported	vertical	vertical	supported	limited
LOTOS [Section 1.5.8]	limited	vertical	limited	limited	supported
MCSE [Section 1.5.9]	limited	limited	supported	limited	supported
ISPME [Section 1.5.10]	supported	supported	supported	supported	supported

Table 3: Support for Model-Continuity Attribute by Specification-Modeling Methodologies

Specification Modeling Methodologies	Language [Section 1.4.1]	Complexity control [Section 1.4.2]	Model continuity [Section 1.4.3]
SDRTS [Section 1.5.1]	limited	limited	limited
drs JSD [Section 1.5.2]	limited	supported	limited
SREM [Section 1.5.3]	supported	limited	limited
rdrl OOA [Section 1.5.4]	limited	supported	limited
SDL [Section 1.5.5]	supported	supported	limited
ECS [Section 1.5.6]	supported	supported	limited
VDM [Section 1.5.7]	supported	limited	limited
LOTOS [Section 1.5.8]	supported	limited	limited
lbrdrl MCSE [Section 1.5.9]	supported	supported	limited
ISPME [Section 1.5.10]	supported	supported	supported

Table 4: Summarizing Attributes of Reactive-System Specification-Modeling Methodologies

1.6.1. Language (Table 1)

There are a number of languages available today that are suitable for expressing reactive system characteristics [4, 5, 23, 26, 28, 43, 59]. However, several of the methodologies we examined did not incorporate such languages. Another interesting feature we observed is that several methodologies support more than one of the following system views: entity, activity, and behavior. For example, ECS, LOTOS, SDL support two of the possible three system views.

Most of the studied methodologies supported explicit models. We do not find this surprising, as it is typically easier, in our view, to conceptualize reactive system behavior in terms of internal states. However, external or property oriented models, such as those created in LOTOS, makes it easier to characterize the system in an implementation-independent way.

Direct support for expressing timing constraints or a model of wall-clock time was not available in several cases. Several methodologies did not directly support an explicit model of time and supported temporal ordering between events instead. Another specification requirement of reactive systems, namely, exception handling, was often not well supported by the specification languages.

Almost none of the approaches allow the characterization of the reactive system's environment. We find that most of the later and developing methodologies adopted languages that allow formal specifications. However, operational semantics is not supported in some cases, especially in the case of LOTOS, where there is no explicit model.

1.6.2. Complexity Control (Table 2)

Representational complexity is generally well supported by the methodologies. In a few cases, neither hierarchy nor orthogonality were supported to the fullest extent. Most methodologies chose visual schemes for representing the specifications, and in some cases such as ECS, associated semantics with the graphical objects that represented the specifications.

Developmental complexity is also reasonably well supported. However, nondeterminism and perfect-synchrony were not well supported in some cases. In some cases, the assumption of perfect synchrony was not made clear. Most formal specification approaches adopted the concept of perfect synchrony.

For guidance in the development of specifications, most methodologies supported the top-down approach. Some allowed an iterative approach, and few supported a bottom up approach. An incremental refinement of the specification was supported in some cases.

1.6.3. Model Continuity (Table 3)

In our opinion, this is the weakest dimension for most methodologies. Except for ISPME[49], none support the integrated modeling dimensions, especially when one considered models that were dissimilar. Implementation-independence of the specification was generally supported. Implementation assistance was not limited in cases where the methodology was non-formal or was based on an external model.

Support for conformance checking, model interaction, or model complexity was limited. We also observed that in some cases the implementations were derived from the structure of the specification. In approaches that support external models, implementations were not usually derived directly from the specifications as the latter had no internal structure. Many of the approaches were also more applicable to software, rather than both hardware and software.

1.6.4. Overall (Table 4)

Overall, we feel that the languages and complexity control dimensions are reasonably well supported today, even though there is room for improvement. Except for ISPME, the model-continuity dimension is not supported well, especially when one considers dissimilar modeling domains.

1.7. RECOMMENDATIONS

Based on the observations above, we make several recommendations regarding what we feel most specification methodologies need to specifically address today. In an overall sense, we think approaches that support all the views, express all reactive system characteristics using a visual formalism, which has an associated formal and operational semantics, would be ideal. Further, this methodology should allow integrated modeling with other models and strongly support implementation independence. Automated generation of possible implementations should also be highly desirable.

We categorize our recommendations according to the dimensions presented earlier.

1.7.1. Specification Languages

In addition to what is supported by the state of the art, specification language(s) adopted by the methodology should incorporate the following guidelines:

1.7.1.1. Conceptual Models

- ***should support entity, behavior and activity views at the same time***
  All three views of a system: entity, behavior and activity, should be supported, since all three views are complementary to each other. Supporting more than one view would possibly entail either supporting more than one specification language, or choosing a language that incorporates more than one view.

- ***should support both model-oriented and property-oriented specifications***
  While it is typically easier to understand and relatively easier to generate implementations from a model-oriented specification, property-oriented specifications are less implementation dependent and offer the implementor to be more creative in generating solutions. We therefore recommend supporting both model-oriented and property-oriented specifications in an integrated manner.

- ***should specify timing constraints explicitly***
  To improve the understandability and to reduce the chances or erroneous specifications, it is preferable to be able to specify timing constraints such as maximum or minimum execution times, data transfer rates, or inter-event constraints directly. We recommend a simple and flexible specification scheme be chosen.

- ***should model passage of time explicitly***
  While temporal ordering may often be sufficient for the purpose of proving properties of the system, it may be easier to understand and express required behavior in terms of explicit time. Therefore, a methodology should model time explicitly.

- ***should support exception handling***
  Exception handling mechanisms should also be provided in the specification language.

- ***should characterize the environment explicitly***
  Some allow the representation of the environment as a separate model. We believe that the environment should be characterized as a property-oriented specification, where the environmental characteristics are specified as properties and hints.

1.7.1.2. Analysis Techniques

- ***should support formal semantics***
  Supporting a specification language with formal semantics will result in unambiguous and rigorously verifiable specifications. Formal semantics will also support automated techniques for checking inconsistencies in the specification and the synthesis and verification of implementations. There has been an increased awareness of the importance of incorporating formal specifications into software development [24]. This trend of awareness should continue to evolve and encompass both software and hardware development.

- ***should support operational semantics***
  By supporting operational semantics, the methodology will enable the generation of early prototypes of the system under design, which would allow early validation of user requirements without necessarily committing to design decisions.

1.7.2. Complexity Control

1.7.2.1. Representational Complexity

- ***should support visual formalisms***
  As pointed out in [28, 29], visual formalisms assist in representing conceptual constructs that specifiers and designers have in mind during the design stages of complex reactive systems.

1.7.2.2. Developmental Complexity

- ***should support bottom-up development***
  We found that the existing specification methodologies rarely allow the composition of specifications in a bottom up manner. We believe that supporting a bottom up approach will assist in specification reuse, where one may potentially create new specifications from a collection of preexisting specifications.

- ***should make explicit assumption of perfect-synchrony hypothesis***
  The assumption of perfect-synchrony has been implicitly adopted by some methodologies. However, this assumption is not always valid, and can cause design scenarios to remain unexplored [49]. The methodology should allow one to explore such cases by explicitly choosing or rejecting the hypothesis.

1.7.3. Model Continuity

1.7.3.1. Integrated Modeling

- ***should support conformance checking between models***
  By supporting conformance between a specification model and models developed during later stages, divergence of these models from specified behavior can be detected.

- ***should allow integrated simulation and formal analysis***
  Currently most methodologies simulate and analyze the specification model in isolation from other models. Simulating and analyzing these models in an integrated manner will result in a synergistic modeling environment. This synergy will be due to an increased interaction between the two models that will result in the exchange of information that may not be available to the models individually. The specification methodologies should exploit simulation languages such as VHDL [36], which are able to represent digital systems at various levels of abstractions and modeling domains.

- ***should allow integrated representation and refinement across different design stages***
  To facilitate a step-wise, iteratively refined design process across all design stages, an integrated representation scheme is needed where one can represent the specification, design, and implementation steps at various stages of development.

1.7.3.2. Implementation Independence

- ***should extend to include both software and hardware***
  Most specifications are either biased towards hardware or towards software. An integrated representation scheme is required.

- ***should allow incorporation of implementation details without sacrificing implementation independence***
  While the specifications are themselves implementation independent, some design processes tend to modify the original specifications by annotating it with implementation dependent information. To compare different implementations for the same specification, one should be able to devise a scheme that allows the inclusion of implementation dependent details in a generic manner. The scheme should support the inclusion of implementation information from different alternatives. For the example of such a technique, see [49].

1.7.3.3. Implementation Assistance

- ***should provide automated support for generating implementations***
  To extend the useful life of the specification, the designers should be able to generate, automatically if possible, reasonably good implementations from the specification.

1.8. CONCLUSIONS

We presented a framework to evaluate specification modeling methodologies for reactive systems. The framework has been synthesized by combining the characteristics of reactive systems and the requirements of a specification modeling methodology. Using this framework, we examine ten methodologies. Such an evaluation technique proved useful, since we were able to study the individual strengths and weaknesses of each methodology in a common-evaluation framework. Further, we were able to make recommendations based on our survey that we believe should be incorporated by both existing and emerging methodologies.

BIBLIOGRAPHY

[1] M. W. Alford, J.P. Ansart, G. Hommel, L. Lamport, B. Liskov, G. P. Mullery, F. B. Schneider. *Distributed Systems. Methods and Tools for Specification.* Lecture Notes in Computer Science, Springer-Verlag 1982.

[2] Aylor, J. H. and Waxman, R. and Johnson, B.W. and R.D.Williams,. The Integration of Performance and Functional Modeling in VHDL. In *Performance and Fault Modeling with VHDL.* Schoen, J. M., Prentice Hall, Englewood Cliffs, NJ 07632, 1992, pages 22-145.

[3] Aylor, J. H. and Williams, R. D. and Waxman, R. and Johnson, B. W. and Blackburn, R. L. *A Fundamental Approach to Uninterpreted/Interpreted Modeling of Digital Systems in a Common Simulation Environment.* UVa Technical Report TR # 900724.0, University of Virginia, Charlottesville, USA, July 24, 1990.

[4] S. Budkowski and P. Dembinski. *An Introduction to Estelle: A Specification Language for Distributed Systems.* Computer Networks and ISDN Systems. North-Holland, Vol. 14, 1987.

[5] G. Berry. *A Hardware implementation of pure Esterel.* Digital Equipment Pars Research Laboratory, July 1991.

[6] Brayton, R. K. and Camposano, R. and De Micheli, G. and Otten, R. H. J. M. and van Eijndhoven, J. T. J. The Yorktown Silicon Compiler System. In *Silicon Compilation.* Gajski, D.D., Addison-Wesley, 1988.

[7] F. Belina, D. Hogrefe, and A. Sarma. *SDL with Applications from Protocol Specifications.* Prentice Hall, 1991.

[8] D. Bjorner, C.B. Jones. *The Vienna Development Method: The Meta-Language.* Lecture Notes in Computer Science. No. 61. Springer-Verlag. 1978.

[9] Blackburn, R. L. and Thomas, D. E. *Linking the Behavioral and Structural Domains of Representation in a Synthesis System.* DAC 85:374-380.

[10] J.P. Calvez. *Embedded Real-Time Systems: A Specification and Design Methodology.* Wiley Series in Software Engineering Practice. 1993.

[11] J.R. Cameron. *"An overview of JSD".* IEEE Transactions on Software Engineering. Vol SE-12. No. 2. February 1986.

[12] CCITT. *Recommendation Z.100: Specification and Description language (SDL).* Volume X, Fascicle X.1, Blue Book, October 1988

[13] D. Coleman. *"Introducing Objectcharts or How to Use Statecharts in Object-Oriented Design"*. IEEE Transactions on Software Engineering. Vol. 18, No. 1. Jan 1992

[14] Chu, C. M. and Potkonjak, M. and Thaler, M. and Rabaey, J. *HYPER: An Interactive Synthesis Environment for High Performance Real Time Applications*. Proceeding of the International Conference on Computer Design, pages 432-435, 1989.

[15] Camposano, R. and Rosenstiel, W. *A Design Environment for the Synthesis of Integrated Circuits*. 11th EUROMICRO Symposium on Microprocessing and Microprogramming 1985.

[16] P. Coad, E. Yourdon. *Object-Oriented Analysis*. Prentice-Hall, 1990.

[17] A. Davis. *"A Comparison of Techniques for the Specification of External System Behavior"*. Communications of the ACM. Vol 31, No. 9. 1988.

[18] T. DeMarco. *Structured Analysis and System Specification*. Yourdon Computing Series, Yourdon Press, Prentice Hall, 1979.

[19] Dutt, N. D. and Gajski, D. D. *Designer Controlled Behavioral Synthesis*. Proceedings of the 26th Design Automation Conference, pages 754-757, 1989.

[20] D. Drusinsky and D. Harel. *Using Statecharts for hardware description and synthesis*. In IEEE Transactions on Computer-Aided Design, 1989.

[21] E.W. Dijkstra. *"Guarded commands, nondeterminacy, and formal derivation of programs"*. Communications of the ACM, Vol 18, No. 8. 1975.

[22] H. Ehrig, B. Mahr. *Fundamentals of Algebraic Specification* - 1. EATCS Monographs on Theoretical Computer Science 6. Springer-Verlag. 1985.

[23] L.M.G. Feijs and H.B.M. Jonkers. *Specification and Design with COLD-K*. LNCS 490, pp. 277-301.

[24] M.D. Fraser, K. Kumar, V.K. Vaishnavi. *"Strategies for Incorporating Formal Specifications in Software Development"*. Communications of the ACM. Vol. 37, No. 10. Oct 1994 p 74-86.

[25] O. Færgemand, A. Olsen. *New Features in SDL-92*. Tutorial, Telecommunications Research Laboratory, TFL, Denmark. 1992.

[26] D.D. Gajski, F. Vahid, and S. Narayan. *A system-design methodology: Executable-specification refinement*. In Proceedings of the European Conference on Design Automation (EDAC), 1994.

[27] N. Halbwachs. *Synchronous Programming of reactive Systems*. Kluwer Academic Publishers, 1993.

[28] D. Harel. *"Biting the Silver Bullet: Toward a Brighter Future for System Development"*. IEEE Computer. Vol. 25, No. 1. Jan 1992, pages 8-24.

[29] D. Harel. *"On Visual Formalisms"*. Communications of the ACM. Vol. 31, No. 5. 1988 p 514-530.

[30] D. Harel. *"Statecharts: A Visual Formalism For Complex Systems"*. Science of Computer Programming, Vol 8. 1987, pages 231-274.

[31] D.C. Hu and G. DeMicheli. *HardwareC - a language for hardware design.* Stanford University, Technical Report CSL-TR-90-419, 1988.

[32] D. Harel, H. Lachover, A. Naamad, A. Pnueli, M. Politi, R. Sherman, A. Shtull-Trauring, M. Trakhtenbrot. *"Statemate: A working environment for the development of complex reactive systems"*. Proceedings of 10th International Conference on Software Engineering. Singapore, 11-15 April 1988, p 122-129.

[33] S. Hekmatpur, D. Ince. *Software Prototyping, Formal Methods, and VDM.* International Computer Science Series, Addison-Wesley Publishing Company, 1988.

[34] C.A.R. Hoare. *Communicating Sequential Processes*. Prentice Hall, London. 1985.

[35] D. J. Hatley, I. A. Pirbhai. *Strategies for Real-time System Specification.* Dorset House Publishing, New York, 1987.

[36] IEEE. *IEEE Standard VHDL Language Reference Manual.* IEEE Inc., NY, 1988.

[37] i-Logix Inc. *ExpressVHDL Documentation*, Version 3.0. 1992.

[38] ISO/IS 8807. *Information Processing Systems - Open Systems Interconnection: LOTOS - A Formal Description Technique*. 1989.

[39] M.A. Jackson. *System Development*. Prentice-Hall, 1983.

[40] M.A. Jackson. *Principles of Program Design*. Academic Press, 1975.

[41] Lor, K. E. and Berry, D. M. *Automatic Synthesis of SARA Design Models from System Requirements*. IEEE Transactions on Software Engineering 17(12):1229-1240 December 1991.

[42] J. Z. Lavi, M. Winokur. *"Embedded computer systems: requirements analysis and specification: An industrial course"*. Proceedings of SEI Conference, Virginia April 1988. Lecture Notes in Computer Science: Software Engineering Education, No. 327. Ed. G. A. Ford, Springer-Verlag p 81-105.

[43] F. Maraninchi. *Argos: A graphical synchronous language for the description of reactive systems.* Report RT-C29, Univeriste Joseph Fourier, 1991.

[44] R. Milner. *A Calculus of Communicating Systems.* Lecture Notes in Computer Science 92. Springer-Verlag. 1980.

[45] D.E. Monarchi, G.I. Puhr. *"A Research Typology for Object-Oriented Ananlysis and Design".* Communications of the ACM. Vol. 35, No. 9. Sep 1992.

[46] Z. Manna, A. Pnueli. *The temporal logic of reactive and concurrent systems: specification.* Berlin Heidelberg New York: Springer. 1991.

[47] J.L. Peterson. *Petri Net Theory and the Modeling of Systems.* Engelwood Cliffs, NJ, Prentice Hall, Inc., New York. 1981.

[48] A. Pnueli. *"Application of temporal logic to the specification and verification of reactive systems."* Current Trends in Concurrency. Lecture Notes in Computer Science. Eds: de Bakker et al. Vol. 224, No. 9. Sep 1992.

[49] A. Sarkar. *An Integrated Specification and Performance Modeling Approach for Digital System Design.* Ph.D. Thesis. University of Virginia. Charlottesville, U.S.A. Feb. 1995.

[50] S. Schlaer, S.J. Mellor. *Object-Oriented Systems Analysis.* Yourdon Press, 1988.

[51] J.M. Spivey. *Understanding Z: A specification Language and its Formal Semantics.* Cambridge University Press. 1988.

[52] R. Saracco, J. Smith, R. Reed. *Telecommunication Systems Engineering using SDL.* Elsevier Science Publishers. 1989.

[53] E. Sternheim, R. Singh, and Y. Trivedi. *Hardware Modeling with Verilog HDL.* Automata Publishing Company, Cupertino, CA, 1990.

[54] A. Sarkar, R. Waxman, and J.P. Cohoon. *System Design Utilizing Integrated Specification and Performance Models.* Proceedings, VHDL International Users Forum, Oakland, California, May 1-4, 1994, pp 90-100.

[55] D.E. Thomas and P. Moorby. *The Verilog Hardware Description Language.* Kluwer Academic Publishers, 1991.

[56] J.M. Wing. *"A specifier's introduction to formal methods".* IEEE Computer, Vol 23, No. 9. 1990. pp 8-24.

[57] Woo, N. and Wolf, W. and Dunlop, *A. Compilation of a single specification into hardware and software.* AT&T Bell Labs, 1992.

[58] P. T. Ward, S. J. Mellor. *Structured Development for Real-time Systems*, Vols 1 & 2. Yourdon Computing Series, Yourdon Press, Prentice Hall 1985.

[59] P. Zave and W. Shell. *Salient features of an executable specification language and its environment*. IEEE Transactions on Software Engineering 12(2):312-325 Feb, 1986.

2

SURVEY ON LANGUAGES FOR OBJECT ORIENTED HARDWARE DESIGN METHODOLOGIES

Guido Schumacher, Wolfgang Nebel

Department of Computer Science, Carl von Ossietzky University Oldenburg, 26111 Oldenburg, Germany

ABSTRACT

A good specification and design of a hardware system should have some principal properties. The specification has to be correct, robust, expandable and reusable. There is a variety of different techniques and methods to solve these problems. This article gives an overview about existing specification languages and techniques which allow one to specify systems at a high level of abstraction. It analyses which possibilities, especially for hardware specification, the language concepts of SDL, Estelle, LOTOS, ML, HML, Z, CSP, OCCAM and some object-oriented extensions of VHDL offer. Emphasis is put on object-oriented methodologies and how they are already realized in these languages. The aim of the object orientation is to provide methods for easy reuse, good maintenance and robustness of specifications. References to existing approaches using these techniques in hardware specification and design systems are given. Several general problems of using some of these languages for hardware specification and design are mentioned. It is explained how specifications described in these languages can be used as a starting point for further design steps. In particular any existing links to the hardware description language VHDL for the synthesis process are clarified.

2.1. INTRODUCTION

The first step in a system development process is the construction of a specification. It describes the system in a formal way at a high level of abstraction. The specification is a point of reference during the design process for the system developers. Therefore, links must be provided to the further design steps. Such a link can be the possibility to simulate parts of the specification together with already synthesized parts of the system in order to validate the synthesis steps. It is important to be able to confirm during each phase of the system development that the requirements of the specification are met. As the requirements may change during the life cycle of a system, the specification technique should allow modifications in a simple and robust way and hence support the cost-efficient maintenance of the system.

In many cases parts of a previously developed system can be reused in a new design. This means that the specification technique must provide methods for the reuse of an old specification and the affiliated system parts which allow easy modifications and adaptations to the new application.

Mainly in the software domain, specification methods were developed to solve these problems. More and more methods were extended by object-oriented techniques.

Although the implementation of a system in hardware differs from one in software, the economical requirements for the development process itself are similar. Hence, proposals were made to adapt techniques from software design and use them for hardware development [10].

The following chapters present an overview of specification techniques and language extensions proposed for hardware system specifications. First, specification techniques are described which originally were used in the software domain and were modified for hardware development. Links to further design steps are mentioned as far as they already exist. Second, approaches to extend the hardware description language VHDL with object-oriented features for specification at a high level of abstraction are reported.

2.2. INTRODUCTION TO OBJECT-ORIENTED DESIGN

Although often used in software development, the term 'object-oriented' hasn't a precise meaning but covers a set of quite different languages, programming techniques and design methods. This is even more the case with hardware design methods. This chapter gives an idea of what is meant by 'object-oriented' design and describes typical features of such techniques.

In object-oriented design the application domain is seen as a set of units communicating with each other. In the design such a unit is modelled by an object. Each object is characterized by the information it can carry. The elements of the objects which store this information are called attributes. The actual information of an object corresponds with its internal state. The information in the attributes can be changed by operations which are made visible by an interface to other objects. These operations are called the methods of an object. In other words, the internal state of an object can be changed by another object only by invoking a method. The communication between objects to invoke a method is called a message. Such a method can change the internal state of the object as mentioned above and in turn can send a message to any other object. The internal values of the attribute are hidden to the other objects. This realizes an effective concept for data encapsulation.

As objects with the same properties often appear in an application domain, the set of such objects plays an important role. Such a set of objects is called a class. A slightly different view is to see a class as a template for the generation of objects with the same properties.

A class in the sense of a template can be reused by defining a new class which takes on all the properties of the old one. The properties of the new class (subclass) can be extended and modified. This builds up a relation between the old and the new class characterized by the common properties.The relation is labelled inheritance. Together with encapsulation, inheritance is one of the most important features of object-oriented design because it is the clue to reuse.

To support the introduced features of object-oriented design, especially the inheritance concept, the object-oriented programming languages offer a technique called polymorphism. That means that a single reference may refer to objects of different classes during the life of a program. Inheritance between the different classes means that different implementations of methods may exist which have the same name. The particular version of the method which is to be invoked as a reaction of a message can only be determined dynamically during runtime and depends on the actual class of the referenced object. As the binding of the message to the particular implementation of the method happens during runtime, it is called dynamic binding.

If the introduced concepts are used and the mentioned techniques applied in a design method, then it can be called object-oriented.

2.3. POSSIBLE APPROACHES

2.3.1. Non-VHDL Approaches

A great number of methods and languages proposed for specification exist. This section analyses the possibilities for hardware specification of the language concepts of SDL, Estelle, LOTOS, ML, HML, Z, CSP and OCCAM with special respect to object-oriented methodologies. Given that there are a lot of other languages, this selection was made to cover a representative spectrum of different specification methods, techniques and formalisms.

2.3.1.1. SDL

Probably the best-known specification language from the telecommunication area is SDL [3] [4]. SDL is a CCITT Standard. It allows specification at a very high level of abstraction. It is based on processes communicating asynchronously by signal routes. The processes are described as statemachines. Instances of the signals are sent between the instances of the processes. These signal instances are queued by the receiver which changes its states depending on the queue-entry. This communication mechanism is a very powerful construct in the language and allows a specification at a high level of abstraction. At the same time, this leads to an overhead and sometimes to an overspecification of the system communication which is unnecessary. To address the needs of easy code reuse in the new version of the language (SDL-93) [4] [15], an inheritance mechanism on types exists called specialization. Types can define statemachines in SDL. The inheritance concept on types therefore allows one to redefine procedures and transitions between states. The language provides an overloading mechanism for this redefined constructs as known from object-oriented languages. Some sort of encapsulation concept is provided by the language with the use of types.

This means that it is possible to specify a system in an object-oriented manner by using SDL-92 at a high level of abstraction. Although the techniques described above are used and therefore the advantages mentioned there are present, there is still one large unresolved problem. There is no link between a SDL description at that high level of abstraction and further synthesis steps. It is not possible, e.g., to simulate synthesized parts of the system within the SDL system description. As synthesis tools do not accept a SDL description to do a synthesis, there has to be a change in the description language and therefore a break in the design methodology while going through the synthesis steps. This is the reason why simulation of synthesized parts within the system description is not possible.

There have been some attempts to solve this problem by building translation tools which translates SDL into a hardware description language accepted by synthesis tools [8][9][10][11] or by simulators [12]. A similar approach is to translate SDL via an intermediate format called SOLAR [14] or HDF [13] to VHDL. But within these approaches there are problems which are still unsolved. The tools do not support the oo-features of the new language definition. The translators which produce VHDL code for further synthesis steps do not support the dynamic language constructs like creation and deletion of a process. The tool which translates into VHDL for simulation purposes [12] accepts the CREATE construct in SDL only if a maximum number of parallel process instances is defined by the user.

Another problem is the modelling of simple communication and synchronisation techniques in SDL. For example, the modelling of a global clock is not provided in the language. The queuing of signals is not appropriate for modelling a clock. In the system described in [14], the translation of the high-level communication constructs of SDL into VHDL is the selection of predefined protocols by the designer and the association of them to SDL signals. Similarly, in another approach [10] [11] libraries of protocols written by the designer are used in RTL or algorithmic specifications to map communication constructs of SDL to VHDL.

As described in [8], in this approach SDL communication constructs are only used for system-level specifications. At lower specification levels the SDL processes are hooked together by schematic entry tools which produce VHDL-netlists as outputs. Only the internal behaviour of each component (entity) is modelled by a SDL process. The method explained in [13] presents a hardware-dataflow model (HDF) to model hardware. The mapping of a HDF description into SDL introduces a special technique for modelling in SDL. In this technique each hardware component is modelled by a SDL process. This process collects the inputs by running through a fixed sequence of states. The output is then calculated in the last state. In this approach there is no longer any correlation between SDL specification and real hardware, e.g. the states of a SDL description of a non-sequential circuit do not have any counterpart in the hardware. As a consequence, today only a subset of SDL not containing the oo-features inheritance and polymorphism and not containing low-level communication mechanisms is supported by existing tools.

2.3.1.2. Estelle

There is another specification language well known in the area of telecommunication which is similar to SDL [7]. This is Estelle [6]. Like SDL, it is based on extended finite state-machines, and there is no appropriate link to the synthesis steps following a system specification in one of these languages. First attempts are made in the COSMOS system [14] for Estelle. But again, this attempt does not cover any oo-techniques like the SDL approach. In addition to the difficulties of translating the dynamic features of the language into another hardware description language, the problem of non-determinism exists, which is a characteristic of Estelle. If there are two transitions from one state into another with the same priority which are executable at the same time, one transition is chosen in a non-deterministic manner by Estelle. Because Estelle is not a simple finite state-machine but an extended finite state machine with variables, it is not possible to apply one of the existing algorithms [18] to translate the non-deterministic finite statemachine into a deterministic one. Therefore, the simulation of an Estelle description causes problems.

Despite the fact that Estelle is an ISO standard, it seems that more and more telecommunication specifiers are better off choosing SDL than Estelle.

2.3.1.3. LOTOS

A third language used in the telecommunication world for describing protocols is LOTOS [5]. In contrast to SDL and Estelle, this language is not based on extended finite statemachines but on process algebra. LOTOS is, like Estelle, non-deterministic and therefore provides powerful constructs to express complex processes, but these constructs can not be simulated. This problem becomes even greater if one tries to construct an automatic translation from LOTOS into a deterministic hardware description language like VHDL.

2.3.1.4. Functional Programming Language

A different approach is to use an existing functional programming language for system specifications. As a system can often be described in a functional way, it seems reasonable to use a functional programming language to specify a system. As in the languages mentioned above, again the missing link between the system level description and the synthesis tools causes difficulties in the design process.

ML

Some work has been done to use the functional language ML [16] [17] for system specification purposes and to integrate it into a design flow. In ML a system is described by recursive function application. In most cases this is done interactively, and the interaction between the ML system and the designer is based on the read-eval-print dialogue, a concept known e.g. from LISP.

The language is strongly typed and supports abstract types. The abstract types form the mechanism for the encapsulation. Although this is a sort of polymorphic type system, it is different from the ones known in the oo-domain. It is possible to write a function which can be either applied to an arbitrary data type or to a specific type. The type of the arguments has to be determined at compile-time, which is different from oo-concepts. ML does not provide an inheritance concept. Even very similar types have to be constructed completely new as different types. For example, in a program it might be useful to bind an integer value to a variable as well as a real number. In that case in ML one has to construct a datatype number with constructors int and float. The meaning is that a variable of type number can either be bound to int or to float. If the original declaration of number has to be extended by adding a constructor complex to number, the datatype has to be redefined, the constructor complex has to be added, and the functions using the datatype number have to be modified.

In contrast to oo-languages, it is not possible in ML to reuse and extend existing code without touching the original code. Constructs to model the timing of a system are missing in ML. Although the language provides some constructs for binding identifiers simultaneously to values, there is no real concept for parallelism. Concurrency cannot be expressed in ML. This makes it more difficult to describe a system consisting of units running almost independently and only communicating sometimes with each other.

HML

The hardware description language HML [19] is an extension of ML. HML has an extended type system compared with ML. The goal is to support the verification process of a system description. The precise semantics of HML, based on the semantics of ML [16], and the strong typing system provides the proof support.

2.3.1.5. Formal Languages: Z

The concept of a functional programming language is extended in the specification language Z [20] by using relations instead of functions. Not only the syntax but also the semantics are defined in a formal way. The meta-language to specify the semantics is based on the Zermelo-Fraenkel axiomatisation of set theory. The formal semantics allows one to prove the correctness of a Z specification in a mathematical way. This is one of the main goals of the language.

The Z notation is a calculus which describes a set, the so-called state space, by using relationships. In difference to set theory, a type concept is added to Z. Operations can be described as working on the state space. Such an operation is a relationship between input variables, output variables and a pair of states, one before and one after the operation [21].

In summary, a Z specification describes a relationship between an initial state and a final state. It is possible to have a relationship between one initial state and several final states; even an infinitive number of states is possible. In other words, Z has non-deterministic operations. This allows one to specify non-computable functions for which finite algorithms to compute the final states do not exist. Therefore, simulators of a Z specification are not available.

Successfully developed hardware starting from a Z specification was designed by refinement of the specification [22]. This refinement was done manually. After each refinement step it was verified that the refinement matches the original specification. The mapping of the final Z specification to the real hardware was done manually. The verification of the refinement steps has to be done by mathematicians who are familiar with the calculus and proof techniques. A concept for reusability does not exist at all in the language.

2.3.1.6. Other System Languages: CSP and OCCAM

In the system described in [14] CSP [23] and OCCAM [24] among other languages should be used for system specification. OCCAM is derived from ideas of CSP. Both are synchronous concurrent languages. They allow the parallel composition of communicating sequential processes. The interaction between the processes is based on unbuffered communication. Input commands address the source, and the corresponding output commands address the destination of the communication. An input or output command is delayed until the corresponding output or input command is executed. It is possible to allow several inputs at the same time; in this case, one input is non-deterministically chosen. This introduces a non-deterministic behaviour in a specification. A timing concept in terms of specifying the execution time of the communication commands is not provided. It is mentioned in [23] that the construction of subroutines as processes operating concurrently with its user process offers the possibility of passing commands to the subroutine. This has the effect of 'multiple entry points' and therefore can be used like a Simula class instance; that is, the language offers an encapsulation mechanism. Further essential concepts of oo-languages like inheritance and polymorphism are not provided. Approaches to apply OCCAM or at least a subset of it for system specification use the language for simulation and as an input for a synthesis [25] [26]. However, the lack of a timing concept in the language restricts the synthesis to delay-insensitive control circuits [25]. This means that the correct operation of the circuit does not depend on any assumptions about delays in the wires or operators of the circuit. Links to other synthesis tools which offer other synthesis techniques are not provided.

2.3.1.7. Classical Programming Languages (C, Pascal, etc.)

When having a look at surveys about design engineers, we may be surprised by the high percentage of them saying they are using C for their system specification. Indeed, programming languages are often used for evaluating such or such an algorithm which will be part of the system. Since these languages only provide sequential programming (no concurrency features are offered) and do not have links to system design methodology, they cannot be considered as specification languages.

2.3.2. VHDL-Based Approaches

Instead of using the languages mentioned in "Non-VHDL Approaches" for system specification and translating them into VHDL, it is possible to use VHDL accomplished by some constructs necessary for efficient system specification as a starting point in a design process. VHDL as it is today does not have any oo-features, which is a requirement for code reuse and maintainability in the lifecycle of a system. Furthermore, the communication has to be described in a very detailed and elaborated manner when specifying a system. This usually causes overspecification. It would be desirable to develop the detailed communication mechanism in a system only during the synthesis steps. Some proposals have been made to extend VHDL in an oo-way. An overview about oo-VHDL extensions is given in [29].

2.3.2.1. Following C++ Philosophy

A proposal presented in [27] is based on C++ constructs which are added to VHDL. The main extension is the presentation of a class concept which introduces a new encapsulation mechanism in the language. As in C++, the class consists of a private section and a public section to determine the visibility of class attributes. Similar to C++, the methods are defined as procedures within the class. The procedures implicitly have access to the attributes. This type of class is called a general class. It can be used to declare individual objects of that class or as a starting point for an inheritance mechanism. Like in C++, it is possible to inherit attributes and procedures.

In contrast to C++, it is possible to add port information to a class. The ports become the interface between an object of that class and the rest of the system. In that case it is necessary to define protocols to perform the methods of the class. The protocols just provide a sequence of stimuli at certain ports. This type is called component class.

In the approach a translation mechanism of the oo-constructs to VHDL is proposed. The main idea is the translation of the general class definition into a record definition. The methods or member functions are translated into functions or procedures with an additional argument of the record type. Access to the attributes are changed into access to the additional argument. Object instantiations are translated to data declarations e.g. global signal declarations. Without giving a detailed algorithm, a translation of a component class is proposed in such a way that the class is translated into a component instantiation. The ports of the component are the ports of the component class. The protocols are converted into procedures. The procedures use a signal connected to the ports as an additional parameter. In both types of classes, polymorphism is not possible. As an example a memory description is given:

Example 1:

```
class ram is
private
  memory : array [1 to n] of byte;
public
  procedure read(address: integer;
                 data   : byte) is
  ...
  procedure write (address: integer;
                   data   : byte) is
  ...
end class
class s_ram : ram is
  ...
```

In the example s_ram is derived from the general class ram and inherits the features from it. The class is translated into a record type and the methods into normal procedures.

Example 2:

```
type ram_record is record
  memory : array [1 to n] of byte;
end record;
procedure ram_read(ram_data : ram_record;
                   address : integer;
                   data : byte) is
 ...
procedure ram_write(ram_data : ram_record;
                    address: integer;
                    data: byte) is
 ...
```

The use of global signals causes problems in that approach. It only works properly without additional requirements if the objects are not used from more than one process. The other restriction is the missing polymorphism for procedures.

To overcome the restriction not to have polymorphism, a proposal was made in [28]. Like in the previous approach, a class construct is added to VHDL. In addition to an inheritance mechanism, it offers concepts for polymorphism. This means that it is possible to reimplement methods. They are marked in the basic class as virtual. During run time it is decided which of several different methods will be executed depending on the class of an object. A proposal for a translation of the extended VHDL to VHDL-1076 1987 is given. It is based on transforming the class definition into a record definition. The methods become procedures with an additional argument of an access type. That access type is a pointer to an object's virtual table. This table is a pointer to the actual object. A direct pointer to the object as additional argument of a procedure is not possible because of the strong type concept of VHDL. As a consequence of using dynamic access types for modelling objects, only variables created by allocators are objects. For the proposed translation of the extended VHDL into VHDL-1076 1987, this means that the global use of objects between processes is not possible because of the missing concept for global variables. The memory example looks similar to the previous one:

Example 3:

```
type memory_type is array(integer range<>) of byte;
class ram is
  variable memory: memory_type(1 to n);
  public procedure read;
  virtual public procedure write;
end ram;
class s_ram
  inherit read from ram
is
  virtual public procedure write;
end s_ram;
```

The translation leads to:

```
package ram is
  type memory_type is array(integer range<> of byte);
  type ram_type is
  record
    memory : memory_type(1 to n);
  end record;
 type pointer_ram_type is access of ram_type;
  type s_ram_type is
    record
      memory : memory_type(1 to n);
    end record;
  type pointer_s_ram_type is access of s_ram_type;
  type ram_virtual_class is (ram_virtual_class, s_ram_virtual_class);
  type p_virtual_table_ram_type is
    record
      ram_virtual_class : pointer_ram_type;
      s_ram_virtual_class : pointer_s_ram_type;
      is_a : ram_virtual_class;
    end record;
  procedure read(oops_p_obj: pointer_ram_type, ...);
  procedure read(oops_p_obj: pointer_s_ram_type, ...);
  procedure write(oops_p_virt: p_virtual_table_ram_type, ...);
end package
```

As can be seen from the example, the translation leads to an enormous overhead, and the result cannot be used as a basis for further synthesis steps because of the dynamic constructs. So this solution is produced only for rapid prototyping of systems without a close link to further synthesis steps.

2.3.2.2. Adding Message-Passing Mechanism

Another approach for rapid prototyping is made in [30]. It focuses on a missing mechanism for message passing in VHDL. It is mainly based on a new construct called EntityObject. It is derived from a VHDL entity. In addition, it has not only generics and ports in its interface but also operations. In the specification they are similar to procedures. As the EntityObjects stand for objects in the oo-sense, the operations are the methods of the object. An inheritance mechanism is introduced on the EntityObjects to address code reuse. In contrast to VHDL, it is no longer necessary to establish a fixed connection between entities or EntityObjects by signals. It is possible to determine the destination of a message during runtime. To achieve this, each EntityObject is given a handle by which it can be addressed. To control concurrency, Ada-like accept and select statements are proposed. They determine whether a message is accepted or put into a message queue for incoming messages. To include this method in a design flow, a translation mechanism for this additional construct and the message-passing mechanism is needed.

Example 4:

```
type memory_type is array ( integer range <>) of byte;
EntityObject ram is
  operation read(address: in integer;
                 data : out byte);
  operation write(address : in integer;
                  data : in byte);
end EntityObject ram;
architecture behaviour of ram is
  instance variable memory : memory_type( 1 to n);
  operation read(address: integer;
                 data: out byte) is
  begin
    data := memory(address);
  end;
  operation write(address: in integer;
                data : out byte) is
  begin
    memory(address) := data;
  end;
end behaviour;
EntityObject s_ram is new ram
end EntityObject s_ram;
```

As mentioned above this proposal is made for rapid prototyping using a fixed message-passing mechanism, and therefore there is no close link to further synthesis steps. The problem is similar to translation problems of the SDL high-level communication mechanisms.

In contrast to the previous proposals, an extension of the language standard is suggested in [31]. Based on a monitor concept proposal for shared variables in VHDL-93, it is suggested to use shared variables as objects in the oo-sense. The access to the variables is done by sub-programs which are related to the type of shared variable. The monitor performs an implicit lock on all in or inout actuals of these sub-programs. The type declaration for the shared variable together with the sub-program declaration is seen as a class declaration. The change from using signals for process communication to using a monitor is a fundamental one. Without further explanations, the extension of this monitor concept to signals, constants and files is suggested. A syntax construct is proposed for introducing multiple inheritance to a class declaration. The variable declarations and sub-program declarations of parents classes can be inherited by a derived class.

Example 5:

```
type memory_type is array (integer range <>) of byte;
type ram is class
  variable memory : memory_type ( 1 to n);
  function read (this: ram; address: integer) return byte;
  procedure write(this: ram; address: integer; data: byte);
end class ram;
subtype s_ram is ram class
  procedure write (this: s_ram; address: integer; data: byte);
end class s_ram;
```

It is a disadvantage of this approach that there is no link to the existing language. To realize the proposals, the language standard has to be changed. It cannot be used just on top of VHDL.

Another extension of the language is proposed in [32] based on derived types and tagged types as they are known from the proposals for the new Ada standard [33]. By adding the word tagged to a record or array definition, it becomes expandable. An inheritance mechanism allows one to add additional record elements to those from the parent's type. Like in Ada9X, class-wide types are introduced by an attribute 'CLASS. As an Ada9X 'CLASS specifies an unconstrained type which is the union of all types derived from the base type, it allows dispatching. By determining the type of an object, a sub-program with an appropriate parameter profile is selected during runtime. Similar to the tagged types, an inheritance mechanism for entities and architectures is proposed. The word tagged is added to an entity declaration. A derived entity then inherits all the features from a parent's entity. It is possible to modify them. New items can be added, or old ones can be deleted.

Example 6:

```
type memory_type is array ( integer range <>) of byte;
type ram is
  tagged record
    memory : memory_type ( 1 to n);
  end record;
type s_ram is derived ram with null;
procedure read(this: ram; ..);
procedure read(this: s_ram; ...);
```

2.3.3. Summary

Comparing the different languages and approaches shows that object-oriented techniques are not yet widespread in the hardware specification domain, although they promise support for correct, robust, expandable and reusable specifications.

The only language of the ones cited above which provides object-oriented techniques and therefore supplies the quality characteristics mentioned is SDL'92. To use it in the hardware design process causes problems because of a missing link from the specification to the synthesis. A translation from SDL into VHDL as a proposed solution does not solve that problem entirely because it is not possible to use the communication part of the SDL specification in a further design step, e.g. the queuing mechanism as a high-level communication construct of SDL is not appropriate for use in the synthesis because general implementation of the queuing mechanism will in most cases result in a much too expensive and inefficient hardware. In the cited approaches these parts therefore have to be replaced by VHDL code. This has to be done manually by using a VHDL library or a schematic entry tool producing VHDL.

At the same time SDL, in contrast to Estelle, LOTOS or Z, does not allow one to explicitly specify non-determinism. However, due to the non-deterministic entry of simultaneous events into queues, its simulation may be non-deterministic.

The use of SDL in the software domain offers the possibility of a common environment for hardware and software specification at a high level of abstraction. The intention of using a language like Z or the related ones ML and HML is to have provable specifications in a very strict mathematical sense. But they do not provide any advantage for the robustness and reuse of a specification. The emphasis of the languages CSP and OCCAM is on the introduction of a communication mechanism for concurrent processes rather than on reusability or robustness. A link from specification to further synthesis steps is not provided.

In contrast, VHDL as a standard language for hardware specification offers the link from synthesis to silicon. Looking at VHDL from the object-oriented point of view, it already has a concept for data encapsulation. But only static overloading as a preliminary stage of polymorphism is part of the language. Polymorphism as it is used in the object-oriented domain is not provided by VHDL. An inheritance concept as an important feature of oo-languages is missing completely in VHDL.

Hence extensions to VHDL as cited can introduce an object-oriented design methodology to hardware specification. Although in these approaches there are still unsolved problems too, it seems that the gap between specification and synthesis is smaller and can be closed. The extension of VHDL in an object-oriented way appears to be an interesting area for further research efforts.

REFERENCES

[1] Girczyc, Emil; Carlson, Steve: *Increasing Design Quality and Engineering Productivity through Design Reuse*. Proceedings 30th. ACM/IEEE Design Automation Conference 1993

[2] Furber, Steve: *Asynchronous Circuits for Low Power*. handouts, tutorial EURO-DAC with EURO-VHDL, Grenoble 1994

[3] CCITT Blue book Recommendation Z.100: *Functional Specification and Description Language SDL*, 1989

[4] CCITT Revised Recomendation Z.100 *CCITT Specification and Description Language (SDL)*, 1992

[5] *LOTOS a Formal Description Technique Based on the Temporal Ordering of Observational Behaviour*. ISO, IS 8807, 1989

[6] International Standard, *Estelle (Formal Description Technique Based on an Extended State Transition Model)* ISO/DIS 9074, 1987

[7] Hogrefe. D.: Estelle, *LOTOS und SDL*. Springer Verlag Berlin,1989

[8] Glunz, W.: *Hardware-Entwurf auf abstrakten Ebenen unter Verwendung von Methoden aus dem Software-Entwurf.* (in german) Dissertation, Fachbereich Mathematik/Informatik der Universität-Gesamthochschule Paderborn, 1994

[9] Glunz, W.; Kruse, T.; Rossel, T.; Monjau, D.: *Integrating SDL and VHDL for System-Level Hardware Design*. Proceedings CHDL'93, Ottawa, Canada 1993

[10] Glunz, W.; Venzl, G.: *Hardware Design Using Case Tools*. Proceedings of the IFIP ,VLSI'91 Conference Edinburgh, 1991

[11] Glunz, W.; Venzl, G: *Using SDL for Hardware Design*. in SDL'91 Evolving Methods; Proceedings of the Fifth SDL Forum Glasgow, North Holland, 1991

[12] Lutter, B.; Glunz, W.; Rammig, F. J.: *Using VHDL for Simulation of SDL Specifications*. Proceedings of the European Design Automation Conference'92, 1992

[13] Pulkkinen, O.; Kronlöf, K.: *Integration of SDL and VHDL for High-Level Digital Design*. Proceedings of the European Design Automation Conference'92, 1992

[14] Jerraya, A. A.; O'Brien, K.; Ben Ismail, T.: *Linking System Design Tools and Hardware Design Tools*. Proceedings CHDL'93, Ottawa, Canada 1993

[15] Færgemand, O.; Olsen, A.: *Introduction to SDL-92*. Computer Networks and ISDN Systems 26 p1143-1167, 1994

[16] Milner, R.; Tofte, M.; Harper, R.: *The Definition of Standard ML*. The MIT Press, 1990

[17] Harper, R.: *Introduction to Standard ML*. School of Computer Science Carnegie Mellon University Pittsburgh, 1990

[18] Hopcroft, J. E.; Ullman, J. D.: *Introduction to Automata Theory, Languages, and Computation*. Addison-Wesley, 1979

[19] O'Leary, J.; Linderman, M.; Leeser, M.; Aagaard, M.: *HML: A Hardware Description Language Based on Standard ML*. Proceedings CHDL'93, Ottawa, Canada 1993

[20] Brien, S.; Nicholls, J.: *Z Base Standard Version 1.0*, 1992

[21] Spivey, J. M.: *The Z Notation A Reference Manual*. Prentice Hall, 1989

[22] Potter, B.; Sinclair, J.; Till, D.: *An Introduction to Formal Specification an Z*. Prentice Hall, 1991

[23] Hoare, C. A. R.: *Communication Sequential Processes*. in Hoare, C. A. R. Hoare; Jones, C. B. (eds): Essays in Computing Science. Prentice Hall, 1989

[24] INMOS Ltd.: *OCCAM Programming Manual*, Prentice Hall International, 1984

[25] Brunvand, E.: *Translating Concurrent Programs into Delay-Insensitive Circuits*. in International Conference on Computer-Aided Design (ICCAD), 1989

[26] Marshall, R. M.: *Automatic Generation of Controller Systems from Control Software*. in International Conference on Computer-Aided Design (ICCAD), 1986

[27] Glunz, W.; Pyttel, A.; Venzl, G.: *System-Level Synthesis in Design*. Michel, P.; Lauther, U.; Duzy, P. (eds.) : The Synthesis Approach to Digital System. Kluwer Academic Publishers, 1992

[28] Zippelius, R.; Müller-Glaser, K.D.: *An Object-Oriented Extension of VHDL*. VHDL-Forum for CAD in Europe, Spring 92 Meeting, 1992

[29] Dunlop, D. D.: *Object-Oriented Extensions to VHDL*. Proccedings of the VHDL International User's Forum, 1994

[30] Covnot, B. M.; Hurst, W. D.; Swamy, S.: *OO-VHDL: An Object Oriented VHDL*. Proccedings of the VHDL International User's Forum, 1994

[31] Willis, J. C.; Bailey, S. A.; Newshutz, R.: *A Proposal for Minimally Extending VHDL to Achieve Data Encapsulation Late Binding and Multiple Inheritance*

[32] Mills, M. T., Lt. Col.: *Proposed Object Oriented Programming (OOP) Enhancements to the Very High Speed Integrated Circuits (VHSIC) Hardware Description Language (VHDL)*. Final Report for 05/04/92-08/04/93 Solid State Electronics Directorate Wright Laboratory Air Force Materiel Command Wright-Patterson Air Force Base, Ohio 45433-7331

[33] ISO/IEC JTC1/SC22 WG9 N 193: *Programming Language Ada, Language and Standard Libraries*, Annotated Draft Version 4.0 IR-MA-1364-3, 1993

3

VSPEC: A DECLARATIVE REQUIREMENTS SPECIFICATION LANGUAGE FOR VHDL

Phillip Baraona, John Penix and Perry Alexander

Departments of Electrical and Computer Engineering,
Knowledge-Based Software Engineering Lab,
The University of Cincinnati, OH USA 45221-0030, USA

ABSTRACT

VHDL allows a designer to describe a digital system by specifying a specific design artifact that implements the desired behavior of the system. However, the operational style used by VHDL forces the designer to make design decisions too early in the design process. In addition, there is no means for specifying non-functional performance constraints such as heat dissipation, propagation delay, clock speed, power consumption and layout area in standard VHDL. Thus, VHDL is not appropriate for high level requirements representation. VSPEC is a Larch interface language for VHDL that solves these problems. VSPEC adds seven clauses to the VHDL entity structure that allow a designer to declaratively describe the data transformation a digital system should perform and performance constraints the system must meet. The designer axiomatically specifies the transformation by defining predicates over entity ports and system state describing input precondition and output postconditions. A constraints section allows the user to specify timing, power, heat, clock speed and layout area constraints. In combination with the architecture declaration, collections of VSPEC specified components can define a high level architecture as interconnected collections of components where requirements of components are known (via a VSPEC description), but implementations are not. This work presents the VSPEC language and associated design methodology.

3.1. INTRODUCTION

VSPEC is a language for declaratively specifying digital systems. It annotates the hardware description language VHDL [10, 16] by adding seven clauses to the entity construct. These clauses allow a digital system to be specified using a declarative style as opposed to the operational style of VHDL. With VHDL alone, the only way to specify a digital system is by describing a specific design artifact that implements the system's desired behavior. On the other hand, VSPEC allows the designer to describe the function of the system without defining the eventual implementation. In short, VSPEC allows the specification of "what" a system should do as opposed to the VHDL description of "how" the system will do it. This is consistent with Hoare's definition of specifications [9].

In addition to allowing the specification of "what" instead of "how", VSPEC addresses another limitation of VHDL: specifying performance constraints. When designing a digital system, meeting certain non-functional (i.e. performance) constraints is equally as important as creating a system that functions properly. A flight control system so slow that it calculates a flight correction after the plane crashes is obviously inadequate. Since they are so important in digital systems, performance constraints should be specified very early in the design process. However, VHDL does not provide a consistent mechanism for specifying these types of constraints. VSPEC addresses this problem by allowing the designer to specify performance constraints such as heat dissipation, propagation delay, clock speed, power consumption and layout area.

VSPEC is a Larch interface language for VHDL. The Larch family of specification languages supports a two-tiered, model-based approach to specifying programs [7]. A Larch specification consists of components written in two languages: an Interface Language and the Larch Shared Language. Interface languages are used to specify the interfaces between program components, including component inputs and outputs as well as the observable behavior of the component. Interface languages exist for a variety of programming languages, including C [6], C++ [2], Modula-3 [12] and Ada [5]. Definitions written in the Larch Shared Language (LSL) are the second component of a Larch specification. LSL is a formal algebraic language that defines the underlying sorts and operators used in the Larch Interface Languages [8, 3].

VSPEC is one part of the COMET research project. The goal of this project is to develop better techniques for rapid prototyping of digital signal processing systems. A detailed description of COMET is beyond the scope of this paper [22], but as the project overview in Figure 1 shows, a COMET user begins by writing a description of the function and constraints of the system in VSPEC. This description is then used to partition the system into hardware and software components with an architecture for connecting these pieces together. Each of these components is synthesized and integrated into a board level implementation of the system that is simulated and verified against the original specification.

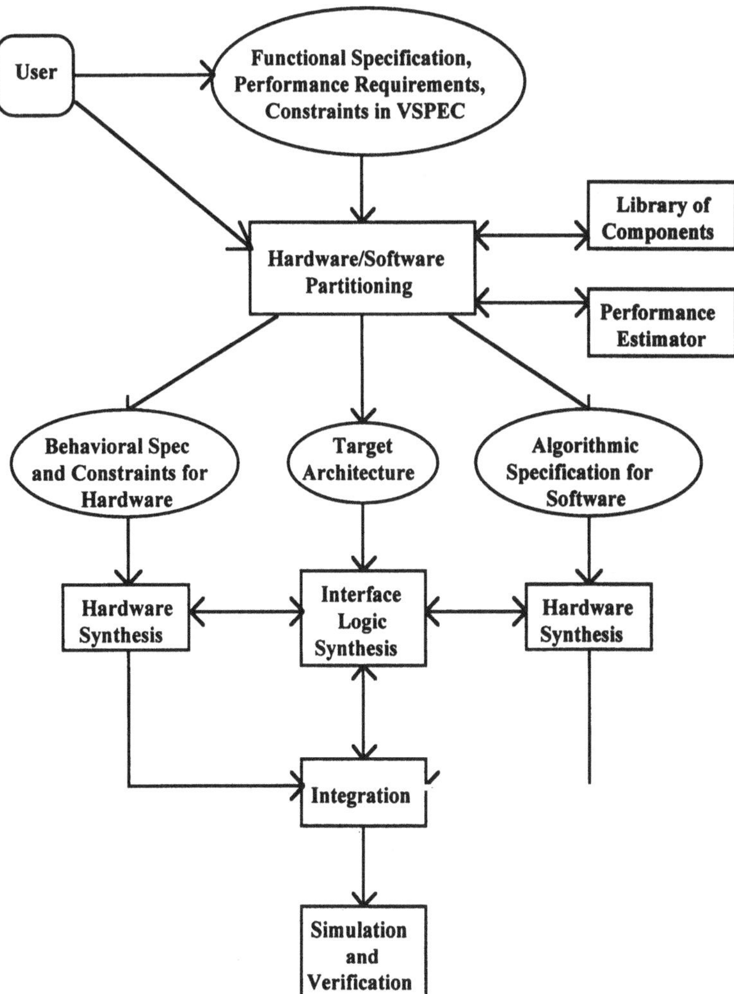

Figure 1: Overview of the COMET Project.

The remainder of this paper gives a more detailed description of VSPEC. The next section briefly describes the VHDL constructs that are important in VSPEC. Section 3.3 gives a detailed description of each of the seven VSPEC clauses while Section 3.4 describes the formal representation of VSPEC. This is followed by an extended example where VSPEC is used to specify a small microprocessor. Section 3.6 discusses other work related to VSPEC and the paper concludes with a description of the current status and future directions for this research.

3.2. IMPORTANT VHDL CONSTRUCTS

This section gives a very brief description of two of the VHDL constructs used in VSPEC. It contains enough information to explain why the VSPEC annotations are needed in a specification language for digital systems. For a complete description of VHDL, refer to the VHDL language reference manual [10] or a textbook on VHDL [16]. If you are already familiar with VHDL, you can skip this section and begin reading about the VSPEC clauses described in Section 3.3.

Two of the more important constructs in VHDL are entities and architectures. A VHDL entity declares a digital component by defining the component's interface. The function of the component is not defined in the entity structure. Instead, each entity has one or more associated architectures where the function of the component is described. This is the "big picture" of how entities and architectures are used. The next few paragraphs give a more detailed description of each of these constructs, starting with the syntax for a VHDL entity:

<entity_declaration> ::= ENTITY *<identifier>* IS
    <entity_header>
    <entity_declarative_part>
    [BEGIN
    <entity_statement_part>]
    END [<Entity_simple_name>] ;

The most important portion of the entity declaration is the entity header. The only part of the entity header currently used in VSPEC is the port clause. A port clause defines the inputs and outputs of the component. Here is an example entity declaration for a simple two input multiplexor:

```
ENTITY vhdl_mux IS
     PORT (D0, D1, cntrl : IN BIT;
            output : OUT BIT);
END vhdl_mux;
```

Notice that this entity merely defines the types of the inputs and outputs to the multiplexor. It does not contain any description of the function of the entity.The function of the multiplexor is described in the VHDL architecture. Each entity has one or more associated architectures. An architecture is used to define the behavior of a specific implementation of an entity. The syntax of the architecture construct is as follows:

<architecture_body> ::= ARCHITECTURE *<identifier>* OF *<Entity_name>* IS
    <architecture_declarative_part>
    BEGIN
    <architecture_statement_part>
    END [*<Architecture_simple_name>*] ;

Detailed descriptions of each portion of the architecture are beyond the scope of this document (see [10, 16]). Suffice it to say that the declarative part of the architecture defines the types, signals and components used by the architecture while the statement part defines the behavior or structure of the entity. Consider the following architecture for the multiplexor entity above:

```
ARCHITECTURE behavior OF vhdl_mux IS
     BEGIN
          PROCESS (D0, D1, cntrl)
               BEGIN
                    IF cntrl = 0 THEN output <= D0;
                    ELSE output <= D1; END IF;
               END PROCESS
END behavior;
```

This is an example of a behavioral architecture. Behavioral architectures use ADA-like programming constructs to define the function of an entity. In this simple example, an if-then statement is used to assign a value (<= is used for signal assignment) to the output port based on the value of `cntrl`. Although this is a simple example, behavioral architectures can be quite complex. Auxiliary procedures and functions can be written in the declarative part of the architecture and entire packages of library routines can be used within the architecture. With these auxiliary procedures and packages, a behavioral architecture can be defined using a large program. No matter what size, all behavioral architectures have one thing in common: they define a single implementation of the behavior of an entity.

Structural architectures are the second common type of VHDL architectures. This architecture type defines the subcomponents an entity is composed of and how those subcomponents are connected. For example, the behavior of the multiplexor could also be defined using and, or and not gates connected as shown in Figure 2. In VHDL, this is represented using the following architecture:

```
ARCHITECTURE structure OF vhdl_mux IS

     COMPONENT and_gate PORT (in1, in2 : IN BIT; output : OUT BIT);
     END COMPONENT;

     COMPONENT or_gate PORT (in1, in2 : IN BIT; output : OUT BIT);
     END COMPONENT;

     COMPONENT not_gate PORT (input : IN BIT; output : OUT BIT);
     END COMPONENT;

     SIGNAL D0_set, D1_set, cntrl_prime : BIT;

BEGIN
     and_1 : and_gate PORT MAP (in1=>D0, in2=>cntrl, output=>D0_set);
     and_2 : and_gate PORT MAP (in1->D1, in2=>cntrl_prime,
                                output=>D1_set);
     not_1 : not_gate PORT MAP (input=>cntrl, output=>cntrl_prime);
     or_1 : or_gate PORT MAP   (in1=>D0_set, in2=>D1_set,
                                output=>output);
END structure;
```

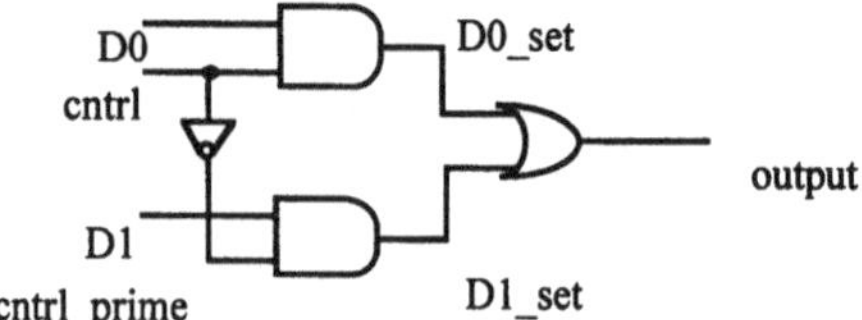

Figure 2: Structural Implementation of Multiplexor.

In this example, the declarative part of the architecture defines three components and three signals. The component declarations (`and_gate`, `or_gate` and `not_gate`) define the inputs and outputs of three sub-components that will be used in this architecture. The behavior and/or structure of these three sub-components must be defined by an entity/architecture pair somewhere else in the system (not shown here). Another VHDL construct, the configuration, is used to map components to the entity/architecture pair that define the behavior of the component. The three signals declared (`D0_set`, `D1_set` and `cntrl_prime`) are used to connect these three components together as shown in Figure 2.

Instances of each of the components in the architecture's declarative part are created in the statement part (between begin and end). The port map for each instance shows how that particular component instance is connected to the signals in the architecture.

Although this example is very small, the same basic concepts defined here scale to much larger systems. This multiplexor could be part of an ALU which is a sub-component of a large microprocessor which is itself one component on a board level system. The same type of structural architecture is used to connect the system together at each of these levels. The lowest level (the and, or and not gates in this example) contains a behavioral description of the components. Because VSPEC is an extension of VHDL, these features for dealing with large systems are also found in VSPEC.

3.3. THE VSPEC CLAUSES

The VSPEC language annotates VHDL by adding seven new clauses to the entity structure. The modified syntax for the entity structure becomes:

<entity_declaration> ::= ENTITY *<identifier>* IS
    <entity_header>
    <vspec_clause_list>
    <entity_declarative_part>
    [BEGIN
    <entity_statement_part>]
    END [*<Entity_simple_name>*] ;

The only change made to the VHDL syntax was the addition of the optional VSPEC clause list to the entity declaration[1]. All other constructs remain intact. A VSPEC clause list is a list of the seven VSPEC clauses separated by semi-colons:

<vspec_clause_list> ::= *<vspec_clause>* { ; *<vspec_clause>* } ;
<vspec_clause> ::= [*<requires_clause>*] | [*<ensures_clause>*] |
    [*<state_clause>*] | [*<constrained_by_clause>*] | [*<modifies_clause>*] |
    [*<based_on_clause>*] | [*<includes_clause>*]

These VSPEC clauses can be grouped into four broad classes. The first class defines the function of the entity and includes the `requires` and `ensures` clauses. The next class declares the internal state of the entity in the `state` clause. The third type of VSPEC clause is used to define the constraints placed on the system. The `constrained by` and `modifies` clauses fall into this category. Finally, the `includes` and `based on` clauses are used to help map the VSPEC definition to its formal representation in LSL. These are the only two clauses that can appear more than once in a VSPEC clause list. The following sub-sections describe each of these clauses in a little bit more detail.

3.3.1. Requires Clause

<requires_clause> ::= REQUIRES *<logical_expression>*

The `requires` clause states the pre-condition for the entity. If the entity's inputs and current state make the `requires` logical expression true, then the entity is guaranteed to perform its specified function. The behavior of the entity is undefined if the `requires` clause is false. A designer that uses an entity specified with VSPEC must ensure that the `requires` logical expression is true before the entity is used. Consider the following example:

```
ENTITY search IS
    PORT (input : IN record_array;
          key : IN INTEGER;
          output : OUT record_type);
    REQUIRES sorted(input);
    ENSURES element_of(output, input) AND output.keyval = key;
    INCLUDES "sorted.lsl", "set.lsl";
END search;
```

In this example, `sorted` is a predicate defined in the file "sorted.lsl" (see description of `includes` clause in Section 3.6) that returns true if the array passed in is in order and false otherwise. The `search` entity above will only function properly if the input array is sorted. If the input is not in order, the function of `search` is undefined. The function of all entities is undefined if the `requires` clause is false. For this reason, it is

[1] This statement is not completely accurate since VHDL's expression syntax was also extended to include quantifiers, logical implication and support for sets and sequences. This is described in a little bit more detail in the VSPEC Language Reference Manual [13].

best to keep the pre-conditions expressed in the `requires` clause as simple as possible. The more conditions that must be met for the `requires` clause to be true (i.e. the more complex the pre-condition), the more difficult it will be to satisfy the pre-condition and use the entity. Thus, the pre-condition should be kept as simple as possible. A pre-condition of true implies the entity has no pre-condition. It must function properly on all input values.

One portion of the `requires` clause definition has been kind of ignored to this point: What is a logical expression? All logical expressions in the VSPEC clauses use a syntax that is an extension of VHDL. The VHDL expression syntax supports the standard boolean expressions and, or and not. VSPEC extends this syntax by adding constructs for variable quantification and logical implication. In addition, the VSPEC expression syntax includes constructs for sets and sequences. See the VSPEC Language Reference Manual [13] for a more detailed description of the syntax of VSPEC expressions.

3.3.2. Ensures Clause

<ensures_clause> ::= ENSURES *<logical_expression>*

The `ensures` clause states the post-condition of the entity. A designer implementing an entity specified with VSPEC must ensure that this logical expression is true whenever the entity processes valid input (i.e. input that makes the `requires` logical expression true). Consider the following example:

```
ENTITY vhdl_mux IS
     PORT (D0, D1, cntrl : IN BIT;
           output : OUT BIT);
     REQUIRES true;
     ENSURES output = (D0 AND cntrl) OR (D1 AND (NOT cntrl));
END vhdl_mux;
```

This is a VSPEC description of the two input multiplexor specified in Section 3.2. The `requires` clause states that this entity is guaranteed to work for all legal values of the input variables. The logical expression in the `ensures` clause declaratively specifies the function of the entity. The logical expression is a condition that must be true when the entity functions properly. Thus, the `ensures` logical expression describes the functional requirements of the entity.

For this simple multiplexor example, the differences between a VHDL behavioral description and VSPEC may not seem that significant. For a more telling example, consider the specification of a sorting component. In VHDL, the simplest way to specify a sorter is an entity with a behavioral architecture describing its function. This behavioral architecture would be an ADA-like description of a specific sorting algorithm such as bubble sort or quicksort. This forces the design of the component to a specific implementation at a very early stage in the design process. In reality, this behavioral architecture is a description of "how" the sorter should work, not "what" the sorter should do. It biases the implementation towards a specific design (i.e. a bubble sort or quicksort) and forces a designer to deal with unnecessary detail at a very early point in the design process.

On the other hand, a sorting component could be described in VSPEC like this:

```
ENTITY sorter IS
     PORT (input : IN ARRAY OF INTEGER;output : OUT ARRAY OF INTEGER);
     REQUIRES true;
     ENSURES permutation(output, input) AND sorted(output);
     INCLUDES "sorted.lsl";
END sorter;
```

In this example, `permutation` is a predicate (defined in "sorted.lsl") that returns true if `output` contains all the same elements as `input` while `sorted` is the same predicate used in Paragraph 3.3.1. This code describes a sorting component as something that ensures `input` and `output` contain the same elements and that `output` is in order. Thus, the specification above describes the functional requirements of the sorter without describing an implementation of a sorting algorithm. In other words, this definition describes "what" the sorter must do instead of defining "how" it should be done. VHDL alone does not allow this type of description. The VSPEC `ensures` and `requires` clause add this feature to VHDL.

3.3.3. State Clause

<state_clause> ::= STATE (*<vspec_variable_declaration_list>*)

The purpose of the `state` clause is to define a list of variables that store the state of an entity. In many algebraic specification languages [7], a computational unit is defined as a transformation from inputs to outputs. This type of transformation is not adequate for specifying systems with VSPEC. Unlike typical subprograms, an entity's local storage is not re-initialized for each use of the entity. Buffers and registers retain their values from one use of the entity to the next. The `state` clause provides a means to model this. The variables declared in the `state` clause serve as the local storage for the entity. In addition, hardware designers very naturally think in terms of the state of a device and the `state` clause allows them to extend this thought process to the specification of the digital system.

The syntax for a VSPEC variable declaration list is:

<vspec_variable_declaration_list> ::= *<vspec_variable_declaration>*
{ ; *<vspec_variable_declaration>* }
<vspec_variable_declaration> ::= *<identifier_list>* : *<subtype_indication>*

An identifier list is a comma-separated list of identifiers while a subtype indication is the VHDL construct used to declare the type of variable. In most cases, this is just an identifier that names the type of the variable(s) declared, but refer to the VHDL documentation for a more complete description [10, 16].

3.3.4. Constrained by Clause

<constrained_by_clause> ::= CONSTRAINED BY *<constraint_logical_expression>*

While the `ensures` clause is used to describe the functional requirements placed on a system, the `constrained by` clause is used to describe the performance requirements of the system. Consider the affect of adding the following clause to the sorter example in Section 3.3.2:

```
CONSTRAINED BY
     size <= 2 um * 5 um AND
     power <= 20 mW AND
     input<->output <= 100 us;
```

With this additional clause, the VSPEC entity now supplies information about the area the entity must be implemented in, the maximum power consumption of the entity and the pin to pin timing for the entity. VHDL does not provide a convenient way to specify these types of performance constraints. The `constrained by` clause provides a standard method for specifying the non-functional requirements of the system.

The logical expression used in the `constrained by` clause must be a conjunction of constraint expressions. The syntax for these expressions is:

<constraint_logical_expression> ::=
<single_constraint> { AND *<single_constraint>* }
<single_constraint> ::= *<constraint_type>* *<relational_op>* *<constraint_value>*
<constraint_type> ::= AREA | HEAT | POWER | CLOCK_FREQUENCY |
<timing_expression>

where the relational operators are the standard VHDL operators <=, <, >=, >, = and /= (not equal) and the constraint value is either a physical literal or a product of two physical literals (i.e. 10 um * 40 um). In VHDL, a physical literal is simply a number followed by a unit (10 mW, for example). Each constraint expression restricts the legal value of the constraint type to a given range, for instance *power* < *1 W*.

VSPEC currently recognizes five constraint types: area, heat dissipation, power consumption, clock frequency and pin to pin timing. In a single constraint, the first four of these constraint types are referenced with the identifier shown in constraint type. A different notation is used to specify the final constraint type, pin to pin timing. The syntax for this type of constraint is:

<timing_expression> ::= *<input_pin>* <-> *<output_pin>*

where input pin and output pin are identifiers that represent an input and an output port of the entity. Thus, an expression such as *input* <-> *output* < 100 *us* states that a change in the data at the input port is propogated to the output port in less than 100 microseconds.

As mentioned above, constraint values are either a physical literal or the product of two physical literals. Area is the only constraint type where a constraint value is the product of two physical literals. Area must be specified in this fashion with the two values representing the bounding box that the entity must fit into. All other constraint types have values that are physical literals.

There are several predefined units that are used for constraint values in VSPEC. The base units of these predefined units are meters for area, watts for power consumption, hertz for clock frequency and seconds for pin to pin timing. In addition to these base units, each of these units can also be expressed using the standard metric prefixes (i.e. area could be fm, um, mm, cm, m or km). VHDL also allows the declaration of virtually any other physical type (see physical type definition in a VHDL reference [10, 16]). In addition to the five pre-defined constraints, VSPEC users can create their own constraint types. At the present time, this has not been implemented in the VSPEC system, but this functionality is a part of the overall plan for the language.

3.3.5. Modifies Clause

<modifies_clause> ::= MODIFIES *<identifier_list>*

The `modifies` clause is used to help build a list of signals and variables the entity will modify. The entity is guaranteed to change only the signals in this modifies list. The value of all other signals in the entity will be left unchanged. Since out mode port signals and all variables in the `state` clause would serve no purpose if the entity did not change them, all out mode port signals and variables in the `state` clause are automatically included in the modifies list. You may explicitly write them in the identifier list in the modifies clause if you desire, but this is an unnecessary step. On the other hand, global variables[1] and buffer/inout mode port signals may only be modified if they are included in the modifies list. It is an error to place in mode port signals in the modifies list since the definition of VHDL does not allow an entity to change the value of an input signal. Here is a simple example to clarify the signals and variables that will and will not occur in the modifies list:

```
ENTITY modifies_example IS PORT (A : IN integer; B : OUT real;
                                 C, D : BUFFER bit; E, F : INOUT bit );
    STATE G : integer;
    MODIFIES C, E;
END modifies_example;
```

The list of signals/variables this entity will modify is `C`, `E`, `B` and `G`. `C` and `E` are included in this list because they are explicitly stated in the `modifies` clause. `B` is included because it is an output signal. All architectures of an entity must assign a value to all entity output signals. Thus, `B` is automatically included in the modifies list. `G` is included in this list for a similar reason. The definition of VSPEC forces the entity to assign a value to all state variables so all state variables are automatically included in the modifies list.

[1] Global variables were added to the 1993 version of VHDL. Previous definitions of the language did not contain global variables.

3.3.6. Includes Clause

<includes_clause> ::= MODIFIES *<string_literal_list>*

The `includes` clause is used to include an LSL description in a VSPEC specification. This LSL description defines the predicates and types used in the specification and it helps map the VSPEC specification to its formal representation in LSL. A VSPEC specification may contain as many `includes` clauses as the user needs to describe the system. We have already seen an example of the `includes` clause in the search entity described in Section 3.3.1:

```
ENTITY search IS
     PORT (input : IN record_array; key : IN integer;
           output : OUT record);
     REQUIRES sorted(input);
     ENSURES element_of(output, input) AND output.keyval = key;
     INCLUDES "sorted.lsl", "set.lsl";
END search;
```

In this example, the file "sorted.lsl" contains the following LSL definition of the `sorted` function:

```
sorted : trait
     includes Integer(integer)
     includes Array1 (record, integer, record_array)
     record tuple of value : anytype, keyval : integer

     introduces sorted : record_array -> bool

     asserts \forall ra : record_array, i : integer
          ((ra[i]).keyval <= (ra[i+1]).keyval)) => sorted(ra)
```

Without going into too many LSL details (see an LSL reference [7, 8, 3]), the definition above states if the `keyval` field of all elements of an array of records is less than or equal to the `keyval` field of the next element in the array, then the array of records is sorted. With this definition, `sorted` becomes a boolean function, or predicate, that can appear in the `requires` or `ensures` clauses to describe a functional requirement of the system. All of the predicates that appear in these clauses must be defined in an LSL file that is listed in one of the `includes` clauses in the entity where the predicate is used.

3.3.7. Based on Clause

<based_on_clause> ::= *<vspec_type>* BASED ON *<lsl_sort>*

The `based` on clause is used to map a data type used in VSPEC to its definition in LSL. This definition in LSL is called a sort. In the syntax above, vspec type is an identifier that refers to the data type used in VSPEC and lsl sort is an identifier that represents the corresponding sort in LSL.

The VSPEC system provides a built in mapping to LSL for all predefined types in VHDL. This is accomplished by automatically including `based on` clauses for these predefined types in all VSPEC entities. The VHDL types `integer`, `boolean` and `string` map to their corresponding types in LSL. The VHDL type `real` maps to the `Rational` sort in the LSL handbook while the `character` type maps to any character in LSL. Finally, the VHDL types `severity_level`, `bit` and `bit_vector` map to the following definitions in LSL:

```
severity_level enumeration of note, warning, error, failure
includes Boolean (bit for boolean, '0' for false, '1' for true)
includes Sequence (bit, bit_vector)
```

This means that the VSPEC systems adds `based on` clauses such as `integer BASED ON Int`, `real BASED ON Rational` and `bit_vector BASED ON bit_vector` to all VSPEC entities. With these clauses included in all VSPEC entities, the predefined types in VHDL may be used in any VSPEC specification.

3.4. FORMAL REPRESENTATION OF VSPEC

All VSPEC definitions can be transformed into a formal definition based on an extension of domain theories defined in CYPRESS [19] and KIDS [21, 20]. CYPRESS and KIDS are software synthesis systems that can be used to generate an executable program from an algebraic specification. A Domain theory describes the problem to be synthesized. It consists of the tuple domain (D), range (R), input pre-condition ($I(x : D)$) and output post-condition ($O(x : D, z : R)$) referred to as a *DRIO* model. In VSPEC, the *DRIO* model can be constructed with these rules:

$D = d_1 \times d_2 \times \ldots \times d_n$ where each d_k is the sort (defined by the `based on` clause) representing the type associated with an `in`, `inout`, or `buffer` port or a state variable.

$R = r_1 \times r_2 \times \ldots \times r_m$ where each r_j is the sort representing the type associated with an element in the modifies list (see Section 3.3.5).

$I(x : D) = I_v(x : D)$ where $I_v(x : D)$ is the logical sentence defined by the `requires` clause.

$O(x : D, z : R) = O_v(x : D, z : R)$ where $O_v(x : D, z : R)$ is the logical sentence defined by the `ensures` clause.

VSPEC is somewhat different from the specification languages that are normally used with CYPRESS and KIDS. A specification language for digital systems must provide a means for describing the performance constraints of the system. One way to do this is to include these types of constraints in the output post-condition of the system. However, this is not the approach taken with VSPEC. Performance constraints have nothing to do with the function of the system so we feel it is appropriate to separate them from the functional requirements defined in the post-condition of the system (i.e. the `ensures` clause). This is one reason the `constrained by` clause is included in VSPEC. The system's performance constraints are specified in the `constrained by` clause while the `ensures` clause describes the functional requirements of the system. The performance constraints can be represented in the formal model of VSPEC by extending the *DRIO* to a *DRIOC* model.

$C(c_1 : C_1, \ldots, c_n : C_n) = C_v(c_1 : C_1, \ldots, c_n : C_n)$ where c_k is a constraint variable such as heat or area, C_k is a sort associated with a constraint variable and C_v is the logical expression defined in the `constrained by` clause.

The definitions in the *DRIOC* describe the system as a transformation mapping the current state and inputs into the next state and outputs such that when the input pre-condition is satisfied the output post-condition and constraints are also satisfied. Formally, this can written as:

$$forall\ x : D\ I(x) \Rightarrow O(x, f(x))\ and\ C(c_1, \ldots, c_n)$$

where $f(x)$ is the transformation performed by the system. This axiom shows the relationship between the design, $f(x)$, and its requirements. In VSPEC, $I(x)$ is derived from the `requires` clause, $O(x,z)$ from the `ensures` clause and $C(c_1, \ldots, c_n)$ from the `constrained by` clause. In VSPEC, $f(x)$ will be defined using behavioral VHDL. Finding $f(x)$ given I, O and C is the synthesis problem addressed by COMET. Proving the equation above is true for a given $f(x)$, I, O and C verifies that $f(x)$ is an implementation of the VSPEC specification.

3.5. EXTENDED EXAMPLE: 16-BIT MOVE MACHINE

3.5.1. Problem Description

The Move Machine is a simple microprocessor whose instructions move data between CPU registers and main memory [18]. The computational units of the machine are assumed to be memory mapped. With this assumption, arithmetic and logical computations are performed as side effects of moving data to and from designated memory locations.

3.5.1.1. Physical Configuration

The physical storage components of the Move Machine are a main memory array and a set of registers. The registers consist of an instruction pointer, an instruction register, and an array of general purpose registers. In this example, a 16-bit Move Machine is specified. The configuration used has 16 general purpose registers, each 16 bits long. The main memory size is 512 bytes (256 16-bit values), requiring 8-bit addressing. The instruction pointer is 8 bits and the instruction register is 16 bits.

3.5.1.2. Instruction Format

The instructions of the 16-bit Move Machine have four fields:

- A two bit op-code. The four operations that the Move Machine has are: load, store, jump, and halt.
- A two bit addressing mode that determines how the effective address is specified in the instruction. The four addressing modes are: absolute, immediate, indirect, relative.
- A four bit register identification to specify which register is to take part in the operation.
- An eight bit effective address which, in conjunction with the addressing mode, determines what memory location takes part in the instruction.

3.5.1.3. Processor Operation

The I/O interface to the Move Machine consists of a start signal, a finished signal and a clear signal. When the start signal is received, the processing cycle begins. When the machine halts (executes a halt instruction), the finished signal is set. The clear signal resets the machine and prepares it to receive the start signal.

The Move Machine uses a three phase processing cycle. In the first phase, the instruction referenced by the instruction pointer is fetched from memory. In the second phase, the effective address is calculated according to the specified addressing mode and the instruction pointer is incremented to reference the next instruction. In the third phase, the fetched instruction is executed.

3.5.2. Specification of the Move Machine

The first step in specifying the behavior of the Move Machine is to define abstract data types in LSL. These types and their associated operations will provide the vocabulary necessary to describe the behavior of the Move Machine. Once this foundation is laid, defining the VSPEC interface specification can begin. First, the input, output, and state variables are specified. Then the desired behavior is described using the appropriate VSPEC clauses.

3.5.2.1. Abstract Types and Operations

Abstract data types and operations are specified using the Larch Shared Language [7]. LSL specifications define operators and sorts that are similar to the programming language concepts of procedure and type. Operators represent total functions from tuples of values to values while sorts stand for disjoint non-empty sets of values. Sorts are used to indicate the domain and ranges of operators.

The trait is the basic unit of specification in LSL. A trait introduces operators and specifies their properties. The trait used to define the operators and sorts for the Move Machine is shown in Figure 3.

```
%% A trait to define the types and operations for the Move Machine.
Move_MC : trait
%% Included traits. Note that bit_vector trait includes bit.
includes Natural (natural)
includes bit_vector (word)
includes Array1 (word, natural, memory_array)
includes Array1 (word, natural, register_array)

%% Define other sorts used in the Move machine.
operation enumeration of load, store, jump, halt
address_mode enumeration of absolute, immediate, indirect, relative
processor_state enumeration of fetch, decode, execute, initial, stop
instruction tuple of op_code : operation,
                     addr_mode : address_mode,
                     reg_id : natural,
                     eff_addr : natural
introduces
    %% Move machine constants.
    word_size : -> Int
    MM_size : -> natural
    OP_size : -> Int
    AM_size : -> Int

 %% Conversion function from word (i.e. bit_vector) to natural.
    word_to_nat : word -> natural
    word_to_nat_with_base : word, natural -> natural

 %% Operations over Move Machine Types.
    decode_op : word -> operation
    decode_am : word -> address_mode
    word_to_instr : word -> instruction

asserts \forall n : natural, w : word, i : instruction
 %% Assign values to constants.
    MM_size = 256; word_size = 16; OP_size = 2; AM_size = 2;
 %% Define the word to natural conversion function.
    word_to_nat(w) == word_to_nat_with_base(w, 1);
    len(w) = 0 => word_to_nat_with_base(w, n) = 0;
    len(w) > 0 => word_to_nat_with_base(w, n) =
             word_to_nat_with_base(substring(w,0,(len(w)-2)), (n*2))
             +(if last(w) = '0' then 0 else n);

 %% Define the decode operations.
len(w) = OP_size /\ w[0] = '0' /\ w[1] = '0' => decode_op(w) = load;
len(w) = OP_size /\ w[0] = '0' /\ w[1] = '1' => decode_op(w) = store;
len(w) = OP_size /\ w[0] = '1' /\ w[1] = '0' => decode_op(w) = jump;
len(w) = OP_size /\ w[0] = '1' /\ w[1] = '1' => decode_op(w) = halt;
len(w) = AM_size /\ w[0] = '0' /\ w[1] = '0' => decode_am(w) =
                                                          absolute;
len(w) = AM_size /\ w[0] = '0' /\ w[1] = '1' => decode_am(w) =
                                                         immediate;
len(w) = AM_size /\ w[0] = '1' /\ w[1] = '0' => decode_am(w) =
                                                          indirect;
len(w) = AM_size /\ w[0] = '1' /\ w[1] = '1' => decode_am(w) =
                                                          relative;
len(w) = word_size => word_to_instr(w) =
    [ decode_op(substring(w,0,1)), decode_am(substring(w,2,3)),
    word_to_nat(substring(w,4,7)), word_to_nat(substring(w,8,15)) ];
```

Figure 3: Move Machine Data Types and Operations.

The trait begins by including a trait called `Natural`. This trait is taken from an LSL handbook. LSL handbooks contain traits that define operators and sorts that can be used in many specifications. This `Natural` trait is a definition of natural numbers and operators over them. It is included here so the specification can use common operators over natural numbers such as subtraction and multiplication.

There are three other traits included in this specification: `bit_vector` and two instance of `Array1`. The `bit_vector` trait defines VHDL bit vectors, while `Array1` is a definition of a one dimensional array. The reason `Array1` is included twice is that all operators and sorts from the `Array1` trait are copied into the `Move_MC` trait and some of the symbols in them are renamed. The `Array1` trait takes three arguments: *E*, *I* and *C*. *E* denotes the sort of the elements of the array, *I* the sort of the array's index and *C* the name of the array. Thus, `Array1` is included twice to create two different types of arrays, a `memory_array` and a `register_array`.

The next code segment creates enumeration sorts for move machine operations, addressing modes and processor states. A tuple is created that corresponds to the instruction format of the Move Machine. Each of these four constructs is created using a shorthand notation in LSL. This shorthand includes a group of operators and sorts that define an enumeration and a tuple. See the LSL documentation [7] for a more accurate description of shorthand notations.

The introduces section lists operators defined by this trait. Each operator's name is shown followed by a map from the list of input sorts to the output sort. Thus, the `word_to_int` operator takes an input of sort `word` and produces an output of sort `natural`. The operators that contain no input sorts declare the sort types of the constants defined by this trait. Values are assigned to these constants in the next section of the trait, the asserts section.

The asserts section contains a list of algebraic equations that define the behavior of the operators in the introduces section. In this example, the first portion of the asserts section assigns values to the four constants in the introduces section. The next section defines what the `word_to_nat` operator does: converts a `word` (i.e. a `bit_vector`) to a `natural`. The behavior of the three decode operators, `decode_op`, `decode_am` and `word_to_instr` are defined in the last section of the trait. The first two of these operators precisely define the operation and addressing mode deciding scheme while the `word_to_instr` operator defines the way that words in memory are decoded.

3.5.2.2. VSPEC Interface Specification

This section contains a detailed description of the interface specification for the Move Machine. The entire specification is shown in Figure 4. We will describe each section of this specification separately, starting with the port declaration. This is where the entity `move_mc` is created and its I/O ports are declared in standard VHDL syntax. The start and clear signals are defined as inputs and the finished signal is defined as an output.

```
entity move_mc is

        port (Start: in bit;            -- Begin processing
              Clear: in bit;            -- Restart processing
              Finished: out bit);       -- Processing completed

includes "Move_MC.lsl";

state    (phase: processor_state;       -- Abstract Processor State
          Memory : memory_array;        -- Main Memory
          IP : natural;                 -- Instruction Pointer
          IR : instruction;             -- Instruction Register
          RGST : register_array;        -- General Purpose Registers
          EA : natural);                -- Effective Address

ensures

     phase = initial implies (Start = '1' implies phase'post = fetch)
          and (Start = '0' implies phase'post = initial)
          and IP'post = 0
          and Memory'post = Memory and RGST'post = RGST
     and  phase = fetch implies IR'post = Word_to_Instr(Memory(IP))
          and phase'post = decode and Memory'post = Memory
          and RGST'post = RGST and IP'post = IP
     and phase = decode implies phase'post = execute
          and (IR.addr_mode = absolute implies
                        EA'post = IR.eff_addr and IP'post = IP + 1)
          and (IR.addr_mode = immediate implies
                               EA'post = IP + 1 and IP'post = IP + 2)
          and (IR.addr_mode = indirect implies
               EA'post = Word_to_Instr(Memory(IR.eff_addr)).eff_addr
               and IP'post = IP + 1)
          and (IR.addr_mode = relative implies
               EA'post = IP + IR.eff_addr and IP'post = IP + 1)
          and  Memory'post = Memory and RGST'post = RGST
               and IR'post = IR
     and phase = execute implies
          (IR.operation = load implies
                                   RGST(IR.reg_id)'post = Memory(EA)
               and forall(x:Register_Id)
                    (x /= IR.reg_id implies RGST(x)'post = RGST(x))
          and (IR.operation /= load implies RGST'post = RGST)
          and (IR.operation = store implies Memory(EA)'post =
               RGST(IR.reg_id)) and forall(x:Address)(x /= EA implies
               Memory(x)'post = Memory(x))
          and (IR.operation /= store implies Memory'post = Memory)
          and (IR.operation = jump implies IP'post = EA)
          and (IR.operation /= jump implies IP'post = IP)
          and (IR.operation = halt implies phase'post = stop)
          and (IR.operation /= halt implies phase'post = fetch))
     and  phase = stop implies Finished'post = '1'
          and (Clear = '0' implies phase'post = stop)
          and (Clear = '1' implies phase'post = initial)
          and Memory'post = Memory and RGST'post = RGST
     and  phase /= stop implies Finished'post = '0';

end move_mc;
```

Figure 4: VSPEC Interface Specification for the Move Machine.

The Move Machine port declaration is:

```
entity move_mc is
    port (Start: in bit;                  -- Begin processing
          Clear: in bit;                  -- Restart processing
          Finished: out bit);             -- Processing completed
```

The VSPEC includes clause follows the port declaration:

```
includes "move_mc_types.lsl";
```

The `includes` clause states that this specification will use abstract types and operations defined in the file `move_mc_types.lsl`, described in the previous section.

The behavior of the Move Machine is specified by describing the allowable transactions between processor states [14]. To do this, we must first define the information that determines the processor state. The Move Machine has a three phase processing cycle that can be viewed as processor states. The addition of a start and a stop state defines a set of states which uniquely describes the status of the Move Machine at any moment in time. The abstract type `processor_state` is defined specifically for this purpose. Therefore, the `state` clause contains the variable `phase` of type `processor_state` to model the processor state:

```
state     (phase: processor_state;     -- Abstract Processor State
          Memory : memory_array;       -- Main Memory
          IP : natural;                -- Instruction Pointer
          IR : instruction;            -- Instruction Register
          RGST : register_array;       -- General Purpose Registers
          EA : natural);               -- Effective Address
```

Naturally, the values of the registers and main memory are of interest when observing the behavior of the processor. Variables of these type are declared in the `state` clause to model these physical structures. Any internal signals that are used to communicate between processor states must be declared as state variables. The effective address is calculated in the decode phase but it is used in the execute phase. Therefore, the variable `EA` of type `natural` is declared to store the effective address between states.

Given a set of input and state variables, the VSPEC `ensures` clause can be used to specify the allowable changes to the output and state variables. In this way, the behavior of the Move Machine is defined. The Move Machine `ensures` clause is structured according to the value of the `phase` variable. This clarifies the specification of the state transactions that are allowed during each phase of processor execution. The allowable transactions for each phase are then conjuncted together to provide a complete behavioral specification.

The permissible next state values must be explicitly constrained for each state variable. If a state variable is not constrained, then presumably it is allowed to take on any value of the associated type. It is not assumed that unconstrained variables remain unchanged. Constraining a variable's behavior is accomplished using the VSPEC `implies` operator to define the next state values that are possible during each processor phase. In this example, the next state values are determinant, but this is not a necessary condition. Non-determinism can be modeled by disjuncting allowable next state values.

The first part of the `ensures` clause specifies what transactions are allowed during the start phase. While in the start phase, the processor is simply waiting for the start signal to begin processing. If the processor does not receive the start signal, it stays in the start phase. This constraint on the next state value of the `phase` variable (`phase'post`) is specified by the first two conjuncts implied by the start phase. Note that the notation `<variable>'post`, where `<variable>` is the identifier for the variable, is used to refer to the value of the variable after the transaction occurs. Here is the part of the `ensures` clause which describes the start phase:

```
phase =   initial implies (Start = '1' implies phase'post = fetch)
          and (Start = '0' implies phase'post = initial)
          and IP'post = 0
          and Memory'post = Memory and RGST'post = RGST
```

The conjunct, `IP'post = 0`, states that the first instruction will be retrieved from memory location 0. The final two conjuncts specify that the main memory and register values must remain unchanged during this processing phase. Without these constraints, the specification would be satisfied by an implementation where the memory and registers values arbitrarily change during this state. Notice that the state variable `EA` is not constrained during this phase. At this point, the `EA` variable does not contain any information which will affect the future state of the machine. Therefore, the specification need not be constrained to retain the value of this variable.

The Move Machine behavior during the fetch phase is described by:

```
phase =   fetch implies IR'post = word_to_instr(Memory(IP))
          and phase'post = decode and Memory'post = Memory
          and RGST'post = RGST and IP'post = IP
```

During the fetch phase, the instruction pointer is updated to contain the interpretation of the word at memory location `IP`. Here, the interpretation is performed by the `word_to_instruction` function defined in the previous section. The next processing phase is specified to be decode, while the memory and remaining register values remain unchanged.

The state changes which occur during the decode phase hinge on the addressing mode. Therefore, the majority of the specification of the decode phase is structured around the value of `IR.addr_mode`:

```
phase =   decode implies phase'post = execute
          and (IR.addr_mode = absolute implies
               EA'post = IR.eff_addr and IP'post = IP + 1)
          and (IR.addr_mode = immediate implies
               EA'post = IP + 1 and IP'post = IP + 2)
          and (IR.addr_mode = indirect implies
               EA'post = word_to_instr(Memory(IR.eff_addr)).eff_addr
               and IP'post = IP + 1)
```

```
        and (IR.addr_mode = relative implies
             EA'post = IP + IR.eff_addr and IP'post = IP + 1)
        and Memory'post = Memory and RGST'post = RGST
                                                    and IR'post = IR
```

The effective address, EA and instruction pointer, IP, are updated according to the current addressing mode. The next phase is specified to be the execute phase. The main memory, the CPU registers and the instruction register are unchanged.

The Move Machine behavior during the execution phase depends upon the fetched operation. This part of the specification is determined by the Move Machine operations:

```
phase = execute implies
        (IR.operation = load implies RGST(IR.reg_id)'post =
        Memory(EA) and forall(x:Register_Id)
        (x /= IR.reg_id implies RGST(x)'post = RGST(x))
        and (IR.operation /= load implies RGST'post = RGST)
        and (IR.operation = store
                      implies Memory(EA)'post = RGST(IR.reg_id))
        and forall(x:Address)(x /= EA
                             implies Memory(x)'post = Memory(x))
        and (IR.operation /= store implies Memory'post = Memory)
        and (IR.operation = jump implies IP'post = EA)
        and (IR.operation /= jump implies IP'post = IP)
        and (IR.operation = halt implies phase'post = stop)
        and (IR.operation /= halt implies phase'post = fetch))
```

For a load operation, the register identified by the current instruction is assigned the value of the memory location referenced by the effective address. This is easily specified by: `RGST(IR.reg_id)'post = Memory(EA)`. However, it is also necessary to specify that the remaining registers do not change. This is the purpose of the second conjunct implied by the load operation. Using the VSPEC `forall` construct, it states that every register that is not involved in the load operation retains its value. When the instruction does not specify a load operation, the values of the register array do not change.

Similarly, for a store operation, the specification states that the specified memory location changes while the rest remain unchanged. The jump operation only effects the value of the instruction pointer. A halt operation causes the next phase to be the stop phase. Any other operation results in the processing returning to the fetch phase.

During the stop phase, the processor sets the finished signal and monitors the clear signal. The stop phase is specified by:

```
phase =   stop implies Finished'post = '1'
          and (Clear = '0' implies phase'post = stop)
          and (Clear = '1' implies phase'post = start)
          and Memory'post = Memory and RGST'post = RGST
and
phase /= stop implies Finished'post = '0';
```

The next phase is determined by the clear signal. This part of the specification also constrains the finished signal to be low during every other phase.The full behavior of the Move Machine is modeled by conjuncting the specifications for the individual phases. Figure 4 shows the entire specification for the Move Machine.

3.6. RELATED WORK

VSPEC uses an axiomatic specification technique common among interface languages in the Larch [7] family of specification languages. Many of VSPEC's constructs are derived from Larch interface languages, most specifically LM3 [12], an interface language for Modula-3. An LSL definition for VSPEC constructs is currently under development. This will facilitate LSL specifications specifically for inclusion in VSPEC constructs.

Odyssey Research Associates (ORA) is also developing a Larch interface language for VHDL [11]. This language differs from VSPEC because it is targeted for formal analysis of a VHDL specification rather than system's level requirements representation. ORA is defining a formal semantics for VHDL using LSL. The LSL representations are used in a traditional theorem prover to verify system correctness. This approach is adopted from the Ada work previously done by ORA in Penelope [4]. Because the ORA work is motivated by the desire to prove VHDL source correct, the interface specification represents proof obligations for the verification process. Thus, the interface language is driven by the need to specify proof obligations for a particular VHDL component at level's of abstraction lower than VSPEC. This is not VSPEC's intent, although doing so is certainly feasible. The purpose of VSPEC is representing requirements and design decisions for classes of solutions at high levels of abstraction.

In ORA's interface language, time is the only non-functional constraint directly represented. Rather than placing constraints on pin-to-pin timing, an absolute time based temporal logic is used to specify an entity's function. This model is based on a general state based approach. One can specify that a predicate *P(x)* must be true in some state *s* using the notation "*P(x) @ s*". Thus, a system's timing constraints are intermingled in the definition of the function of the system. The VSPEC notation specifies time intervals as constraints independent of system function. In principle, separation of concerns is a design goal for any specification language. In practice, including temporal aspects in the functional specification requires use of theorem provers capable of temporal reasoning. Currently, there are few such production quality provers. In VSPEC, information needed for constraint verification is included, but one may choose characteristics for verification.

VAL [1] is another attempt to annotate VHDL. VAL (**V**HDL **A**nnotation **L**anguage) is based on similar work done with Anna for Ada programs [15]. VAL differs from VSPEC because it is an annotation of a specific VHDL design rather than a representation of the requirements for a system not yet designed. VSPEC clauses may access only ports defined by the entity and variables defined locally in the specification.

VAL annotations exist throughout the VHDL specification and formally document its behavior. Any local variable may be referenced in a VAL annotation. Specific aspects of both the structural and behavioral implementation are documented in the VAL annotation. VAL's intent is to document a design for verification where VSPEC's intent is to define requirements for a system.

3.7. CURRENT STATUS AND FUTURE DIRECTIONS

Current VSPEC research involves pursuing domain specific support for prototype synthesis. The role of VSPEC in the COMET system is as a requirements specification language and as input to synthesis tools. Thus, we are working to develop techniques to transform VSPEC into behavioral and structural VHDL. An important related technology transfer issue is developing a handbook of reusable specifications. In the Larch tradition, a handbook is simply a collection of reusable theories defined in the shared language. Handbook theories represent commonly used structures, algorithms and characteristics as well as domain specific information. For VHDL, theories representing standard VHDL types, low level logic functions, signal attributes and conversion routines are some libraries currently being implemented. Theories for pin-to-pin timing, heat dissipation, power consumption, area and clock speed have been implemented to support constraint checking during the design process.

A formal VSPEC definition is under development to precisely define: (a) the relationship between `requires` and `ensures` clauses and simulation; and (b) composition of VSPEC components. Although the abstract notions of pre- and post-conditions are well understood, specific relationships must be established with VHDL's simulation semantics. Using structural VHDL to compose VSPEC is a convenient way to compose VSPEC annotated entities. It is necessary to precisely define the semantics of VSPEC composition in this manner. As with pre- and post-conditions, the meaning of VSPEC composition must be defined with respect to the simulation semantics of VHDL.

A VSPEC parser and type checker has been developed and will be used to drive synthesis tools and the translation from VSPEC to LSL. The parser is developed using the Software Refinery [17]. The Software Refinery provides a suite of program transformation tools. The Dialect tool was used to create a parser that parses VHDL-93 with VSPEC extensions into an abstract syntax tree. This data structure serves as the basis for interfacing VSPEC with other tools.

3.7.1. Acknowledgments

Support for this work was provided in part by the Advanced Research Projects Agency and monitored by Wright Labs under the RASSP Technology Program, contract number F33615-93-C-1316. The authors wish to thank Wright Labs and ARPA for their continuing support and direction. In addition, the authors would like to thank Paul Bailor, Philip A. Wilsey, Hal Carter and Ranga Vemuri for their continuing comments and criticisms of VSPEC.

REFERENCES

[1] L. Augustin, D. Luckham, B. Gennart, Y. Huh, and A. Stanculescu. *Hardware Design and Simulation in VAL/VHDL*. Kluwer Academic Publishers, Boston, MA, 1991.

[2] Yoonsik Cheon and Gary T. Leavens. *A Quick Overview of Larch/C++*. Journal of Object-Oriented Programming, 7(6):39-49, October 1994.

[3] Stephen J. Garland, John V. Guttag, and James J. Horning. *Debugging Larch Shared Language Specifications*. Technical Report 60, Digital Equipment Corporation Systems Research Center, 130 Lytton Avenue, Palo Alto, CA 94301, July 1990.

[4] David Guaspari. Penelope, an Ada Verification System. In *Proceedings of Tri-Ada '89,* pages 216-224, Pittsburgh, PA, October 1989.

[5] David Guaspari, Carla Marceau, and Wolfgang Polak. *Formal Verification of Ada Programs*. IEEE Transactions on Software Engineering, 16(9):1058-1075, September 1990.

[6] John V. Guttag and James J. Horning. Introduction to LCL, *A Larch/C Interface Language*. Technical Report 74, Digital Equipment Corporation Systems Research Center, 130 Lytton Avenue, Palo Alto, CA 94301, July 1991.

[7] John V. Guttag and James J. Horning. *Larch: Languages and Tools for Formal Specification*. Springer-Verlag, New York, NY, 1993.

[8] John V. Guttag, James J. Horning, and Andres Modet. *Report on the Larch Shared Language: Version 2.3*. Technical Report 58, Digital Equipment Corporation Systems Research Center,130 Lytton Avenue, Palo Alto, CA 94301, April 1990.

[9] C.A.R. Hoare. *Algebra and Models*. Proceedings of the First ACM SIGSOFT Symposium on the Foundations of Software Engineering, 18(5):1-8, December 1993.

[10] Institute of Electrical and Electronics Engineers, Inc., 345 East 47th St., New York, NY 10017. *VHDL Language Reference Manual*, 1994.

[11] D. Jamsek and M. Bickford. *Formal Verification of VHDL Models*. Technical Report RL-TR-94-3, Rome Laboratory, Griffiss Air Force Base, NY, March 1994.

[12] Kevin D. Jones. LM3: *A Larch Interface Language for Modula-3, A Definition and Introduction: Version 1.0.* Technical Report 72, Digital Equipment Corporation Systems Research Center, 130 Lytton Avenue, Palo Alto, CA 94301, June 1991.

[13] Knowledge Based Software Engineering Laboratory, University of Cincinnati. *VSPEC Language Reference Manual*, 1995. In Preparation.

[14] Leslie Lamport. *A Simple Approach to Specifying Concurrent Systems*. Communications of the ACM, 32(1):32-45, January 1989.

[15] D. Luckham and F. von Henke. *An Overview of Anna, a Specification Language for Ada*. IEEE Software, 2(2):9-22, March 1985.

[16] Douglas L. Perry. *VHDL*. McGraw-Hill, Inc., New York, NY, 1991.

[17] Reasoning Systems Inc., Palo Alto, CA. *Refine User's Guide*, Version 3.0, May 1990.

[18] Jayanta Roy, Nand Kumar, Rajiv Dutta, and Ranga Vemuri. *DSS: A Distributed High-Level Synthesis System*. IEEE Design&Test of Computers, pages 18-32, June 1992.

[19] D. Smith. *Top-down Synthesis of Divide-and-Conquer Algorithms*. Artificial Intelligence, 27(1):43-96, Sept. 1985.

[20] D. Smith. *Algorithm Theories and Design Tactics*. Science of Computer Programming, 14:305-321, 1990

[21] D. Smith. *KIDS: A Semiautomatic Program Development System*. IEEE Transactions on Software Engineering, 16(9):1024-1043, Sept. 1990.

[22] R. Vemuri, H. Carter, and P. Alexander. *Board and MCM Level Synthesis for Embedded Systems: The COMET Cosynthesis Environment*. In Proceedings of the First Annual RASSP Conference, pages 124-133, Arlington, VA, August 15-18, 1994.

4

COMMUNICATION PROTOCOLS IMPLEMENTED IN HARDWARE: VHDL GENERATION FROM ESTELLE[1]

Jacek Wytrebowicz, Stanislaw Budkowski

Institut National des Télécommunications, 9 rue Charles Fourier, 91011 EVRY, France

ABSTRACT

In this paper we present a method of translation from a specification written in the ISO standardized Estelle language to the standard hardware description language VHDL. The objective is the rapid hardware prototyping of communication protocols. The Estelle formal description technique is used for specification and validation of communication protocols. VHDL is considered an intermediate step, taking advantage of the existing simulation and synthesis tools.

4.1. INTRODUCTION

The complexity of modern day communication protocols is growing to supply intelligent services and to provide higher speed and reliability. Designers specify protocols in a formal way to avoid ambiguity and to be able to validate them. They need methods and tools for the generation of fast prototype implementations, starting from previously validated formal specifications.

[1] This work was, in part, supported by CNET-France Telecom (93PE7404).

A formal description technique Estelle [1] [5] is an international standard since 1989. Estelle is considered as an implementation-oriented formal description technique. VHDL is the only standard hardware description language [4]. The VHDL language does not have formally defined semantics, in contrast to Estelle language. For both Estelle and VHDL, many commercial and public domain tools exist. Among them are editing facilities, translators (which perform lexical, syntactical, and static semantics analysis), debuggers, simulators, C/C++ generators, test generators, and for VHDL only: logic synthesis tools, functional abstractors, behavior/data_flow comparators, and even high level synthesis tools.

In this paper we will concentrate on the translation method from Estelle[1] to VHDL. The translation is a complex task, and some implementation decisions have to be taken to transform highly abstract Estelle specification to a VHDL architecture. Moreover, some Estelle restrictions must be accepted because several Estelle semantic notions cannot be translated. The restrictions are applied to nondeterministic behavior and to dynamic object creation and linking (the hardware components and links are static).

This paper discusses the semantic elements of the two languages and a translation model from Estelle to VHDL. An Estelle to VHDL translator is designed in compliance with the translation model presented here. All example translation results, which are included in the appendix, are extracted from our translator output texts.

4.2. VHDL VERSUS ESTELLE SEMANTICS

4.2.1. Estelle Semantic Model

Estelle is a language for specifying distributed or concurrent processing systems, in particular those that implement protocols and OSI services. The Estelle semantics is defined in a formal way using rigorous mathematical notation.

An Estelle specification describes architectural and behavioral characteristics of a given system. The basic architectural components are module instances. Modules can be nested to create a hierarchical structure. The most external module is called a *specification* module. A module behavior is represented by an extended finite state automaton. Protocol specifications are based on a finite state automation model. Estelle enhances the finite state model by adding control structures and Pascal data types.

The modules may be attributed *systemprocess*, *systemactivity*, *process*, or *activity*. An attribute of a module determines the synchronization between the modules' children. System modules (those attributed *systemprocess* and *systemactivity*) work independently, i.e., asynchronously. Children of *process* and of *systemprocess* modules work in parallel, i.e., their transitions are executed in parallel, and all of them have to terminate to proceed to the next choice and parallel execution of a new set of transitions.

[1] Similar works have been started for other formal description languages, i.e., LOTOS [6] and SDL [3].

Children of *activity* and of *systemactivity* modules work in a nondeterministic way - after a transition is terminated in a child module, a transition at any child module may be executed next.

A parent module has priority over his children. This means that, if a parent module offers a transition, then this transition will be executed regardless of the child's possible transitions offered for execution. This excludes any parallelism among instances which are in an ancestor/descendant relationship. Figure 1 shows an example hierarchy of modules and a snapshot of their work activity.

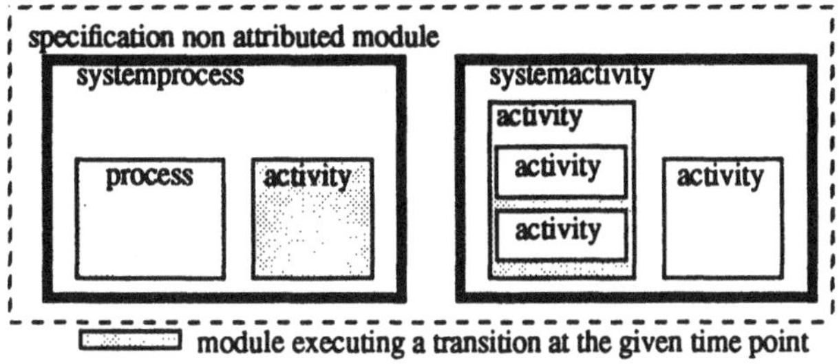

Figure 1: Example of an Estelle System Snapshot

A system module's internal architecture and connection links may be dynamically re-configured. A module transition may contain the instruction to dynamically create or destroy the module children and change interconnections between those existing.

The main communication mean between modules consists in exchanging messages through connected interaction points. Only point-to-point connections are allowed. Another communication mean is to share variables, restricted to modules, which are in the parent-child relation.

A channel declaration describes the messages allowed to be sent and received by each oftwo connected points. A message may carry a set of parameters. Messages are received by interaction points and are stored in unbounded FIFO queues. A queue is attached to one (*individual queue* attribute) or many (*common queue* attribute) interaction points. An interaction point declaration indicates: queue attribute, channel, and a role this interaction point will play in the channel. Figure 2 shows an example of a channel and interaction points declarations.

There are two categories of interaction points:

- external - they are visible to the owner and its parent module,
- internal - they are visible only to the owner module.

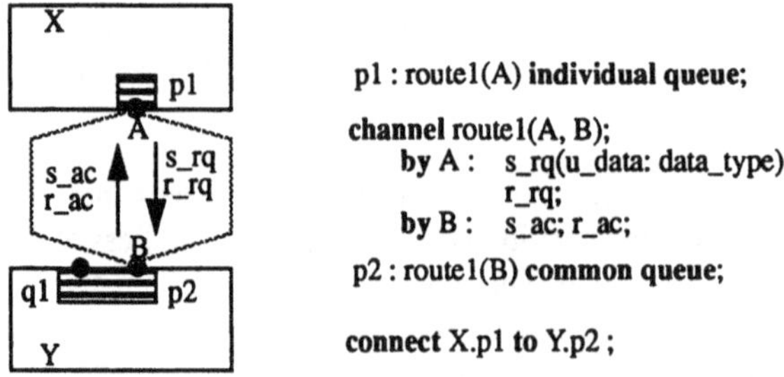

Figure 2: Example of an Interaction Point Link

There are two kinds of interaction point bindings: connections and attachments, as Figure 3 illustrates. An attachment can be perceived as an external interaction point readdressing to a child module to serve this point.

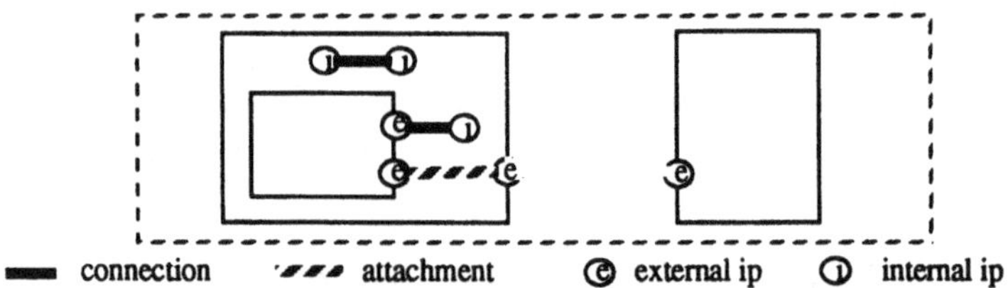

Figure 3: Links of Internal and External Interaction Points

A module instance is a process that performs two actions: transition selection and transition execution. Transition execution is an atomic computation, which means that during its execution, there are no observable changes of the module state, and that a fired action must be completed before the start of another action in this module. A transition execution is a sequential computation of statements that are given in this transition. A selection step is the evaluation of all transition clauses in the module. A transition clause may check:

- automaton (module) actual state - *from* clause,
- presence of a specified message at the top of a given input queue - *when* clause,
- state of local variables, shared variables, or message parameters - *provided* clause,
- time delay for a given transition - *delay* clause,
- priority of the transition with respect to others - *priority* clause.

A transition execution may:

- change the automaton state,
- send a message,
- change local variables,
- create (or kill) a module instance,
- create (or remove) a local link (connection or attachment).

A transition which has all its clauses satisfied is called firable. A module selects one of its firable transitions for execution. Figure 4 shows an example snapshot of a module transition state.

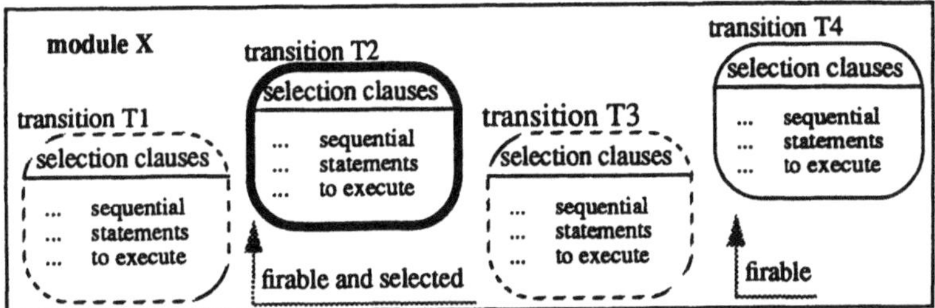

Figure 4: Example of a Module Transition State

The computational model of Estelle is expressed in time-independent terms. It is considered that the selection and execution times of a transition are implementation dependent. The only time dependent statement is a *delay(E1, E2)* clause. The intention of it is to specify that a transition execution must be delayed at least a given time (E1), and that after being delayed continuously for E2 time, the transition will be considered as all other firable transitions. The only assumptions about time in this semantic model are that:

- time progresses as computation does, and
- this progression is uniform with respect to delay values.

A transition execution may lead to data passing through an interaction point or via an exported variable. All passing data reach their destinations before or at the end of a transition execution. The following selection step is performed at the new stable state of the module. Therefore, one can consider the data transmission time as zero, or that data transmission time is included in transition execution time.

There are two nondeterministic steps in the computational model of Estelle:

- An *activity* (*systemactivity*) module makes the nondeterministic choice of a child module to work.
- A module makes the nondeterministic choice of a transition to execute among all firable transitions.

The nondeterminism may model some realizations of a specified system. Any of these realizations works in a deterministic way. During specification the concrete implementation is not known, and thus a class of realizations should be described. For analysis purposes a system under design is specified as well as its environment. Frequently, environment actions have random character, and therefore the nondeterminism may be used to model them.

In some cases, the nondeterministic choice of activating a child module to work may emulate the action of parallel processes. This interpretation is correct in the Estelle specification, when there is no *priority* clause being used together with *when* clause, and also there is no dynamic creation and release of modules and communication links [1].

4.2.2. VHDL Semantic Model

VHDL is a general hardware description language. The VHDL semantics is defined in terms of the operation of an abstract simulator and, therefore, in terms of the models upon which the abstract simulator operates.

The basic notions of VHDL are entity instances and signals. An entity may represent any piece of hardware such as a functional block, a printed circuit board, an integrated circuit, and a macro cell, or even a whole system. A VHDL specification is composed of a set of independently defined entities.

Signals are composites of a value of a given type, the history, and planned future. It is possible to access the current value of a signal as well as the value this signal had at a given time in the past, e.g.:

Example 1: signal_S1 <= signal_S2 + 5;
Example 2: IF signal_S4'DELAYED(100 ns) = 0 THEN

All assignments in which the signal is at the left-hand side are considered as planned values for now or for a given moment in the future, e.g.:

Example 3: signal_S6 <= 0 AFTER 10 ns;

A signal may be perceived as a shared variable, which stores both values of a given type and their realization date.

An entity may have a hierarchical structure, i.e., may be indirectly composed of other entities, as Figure 5 shows. The direct structural composites are components. A configuration statement binds a component instance with an entity. The hierarchical structure of entities is used to decompose a complex hardware into manageable pieces. All of them work in parallel, the structure does not express synchronization dependences between entities. The entity instances are static elements.

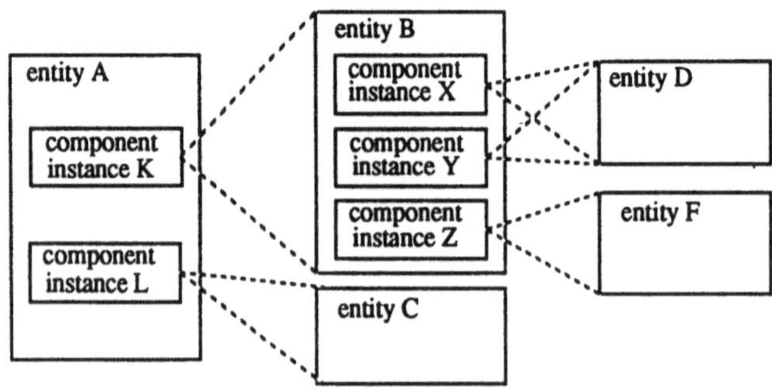

Figure 5: Example of a VHDL Hierarchical Structure

The only means of communication between one entity and another, or between an entity and its parent entity, are signals. Any entity offers some signals for external access. These signals are given within an entity (and a component) port list declaration. An entity associates its local or port signals with its components' (its child entities') port signals. Figure 6 shows an example of signal connections.

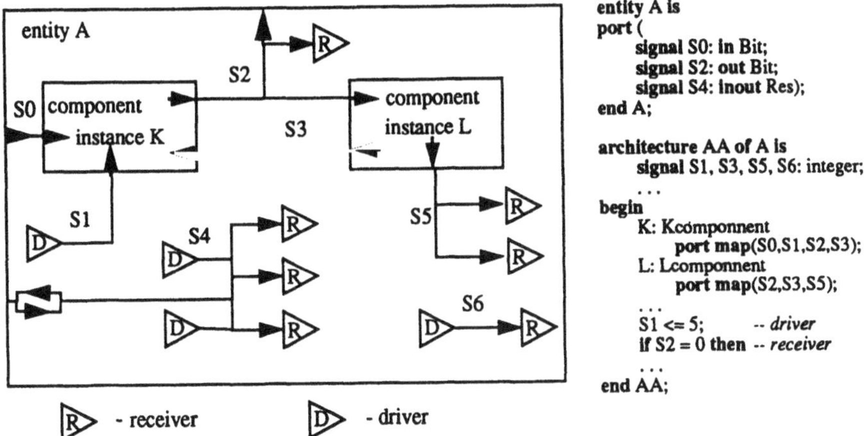

Figure 6: Example of a VHDL Signal Connections

Usually a signal connects one driver with multiple receivers. A driver is instantiated by a VHDL concurrent statement, which description contains an assignment, whose left hand side is a signal (see Example 1 and Example 3). A receiver is instantiated by a VHDL concurrent statement, which tests this signal (see Example 1 and Example 2). A signal may have more than one driver (e.g., S4 signal on Figure 6). A resolution function should be defined for such a signal. This mechanism allows the description of wire-or connections and similar.

An entity behavior is specified by a set of concurrent statements. Among VHDL concurrent statements, there is a *process* statement which embodies an explicitly written sequential program. Such a program can have its local variables. Any concurrent statement is an action, which can be described by a sequential program (an equivalent *process* statement can be written). The exception to this rule are three types of VHDL concurrent statements (i.e., *block* statement, *component instantiation* statement, and generate statement). Any of the three statements can be replaced by a set of processes. Putting aside the syntax details, we can say that an entity behavior is defined by a given number of processes. Figure 7 illustrates a behavioral model of a VHDL specification. A process is executed cyclically each time its enable conditions are satisfied. A process execution is an atomic action. An enable condition may check:

- value change of a given signal at the last simulation step - *wait on* statement (sensitivity clause),
- state of some signals or variables - *wait until* statement (condition clause), and
- time delay - *wait for* statement (time-out clause).

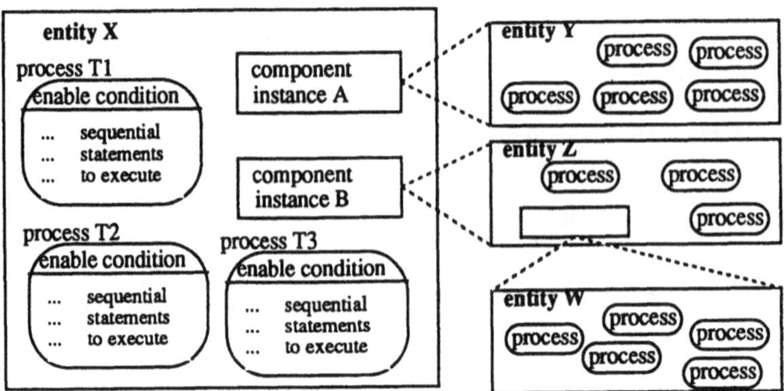

Figure 7: Computation View on VHDL Specification

The abstract simulator evaluates all processes concurrently, regardless of their place in the specification hierarchy. The simulator works in sequential steps of delta time duration. The delta time has a zero value for simulation time counting. The simulation time is a real time model in which one can observe a specified device behavior. The simulator selects all processes which are ready to execute at the current simulation time. If there are no such processes, the simulation time is incremented. Otherwise, the simulator performs one step - it executes all enabled processes in parallel or in some order.

The processes communicate only by signals. Designer defines the delay time of a new signal value setting explicitly in a signal assignment statement. The default value of this time is delta time, which implies that the new signal value will be visible in the next simulation step.

4.2.3. View of Estelle Items in VHDL Terms

Both an Estelle specification and a VHDL specification may describe a designed system together with its environment. They may, however, also describe only a part of a designed system. The modules' hierarchy corresponds in a natural way to the functional decomposition of a system. Some of the system functions (modules) can be implemented in a hardware. A designer may thus select which Estelle modules should be translated to VHDL. The selection of a module implies that all its descendant modules are also selected.

There are some obvious mappings between Estelle and VHDL, as Figure 8 shows. A VHDL entity corresponds to an Estelle module. A component instance represents a child module instance. A configuration statement gives the binding between the component instance and an entity. An Estelle module definition is composed of a module header declaration (external view) and of a module body declaration (internal view). The external view of an Estelle module can be translated to the external view of a VHDL entity, i.e., the entity declaration. The internal view of a module can be translated to the internal view of an entity, i.e., architecture declaration. The Estelle specification module does not have an external visibility, and consequently the corresponding entity declaration is empty.

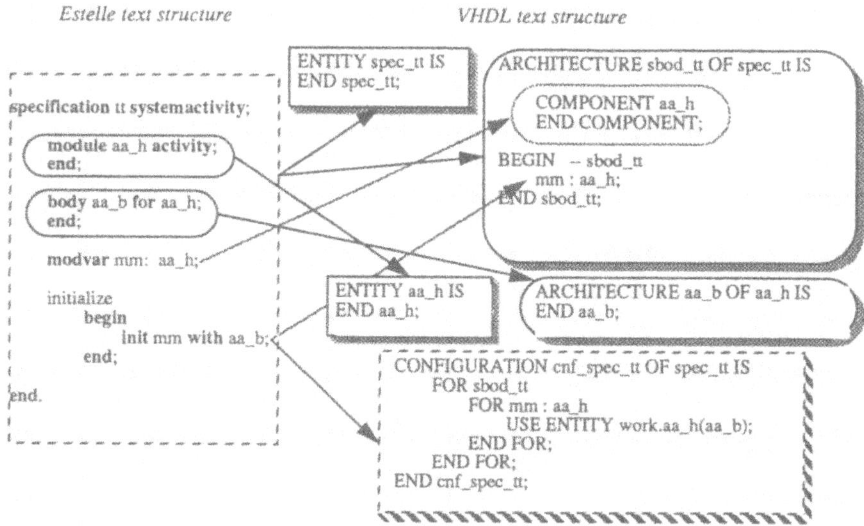

Figure 8: Structure Correspondences between Estelle and VHDL

An Estelle module performs a sequential process, which is composed of an initialization step and an infinite loop of transition selection and transition execution steps. Such a process can be expressed in VHDL as a *process* statement.

A transition selection step may give the control to a child module (or modules), if there is no firable transition within this module. The synchronization, which is expressed by module attributes and nesting, can be translated into VHDL. A solution to reach this goal was considered. It consists of three additional signals between any child entity and its parent one, e.g.:

```
PORT (    -- some signals       -- which correspond to interaction
                                -- points and exported variables
          run : IN boolean;     -- enables computation which
                                -- corresponds to a transition execution
          live : OUT boolean;           -- states that the computation is
                                        -- under execution
          firable : OUT boolean);       -- states that there is
                                        -- a computation to execute
```

These signals will be used to realize synchronization. Figure 9 shows possible waveforms of the synchronization signals. This synchronization mechanism is repeated in all the nested entities, which correspond to the nested modules.

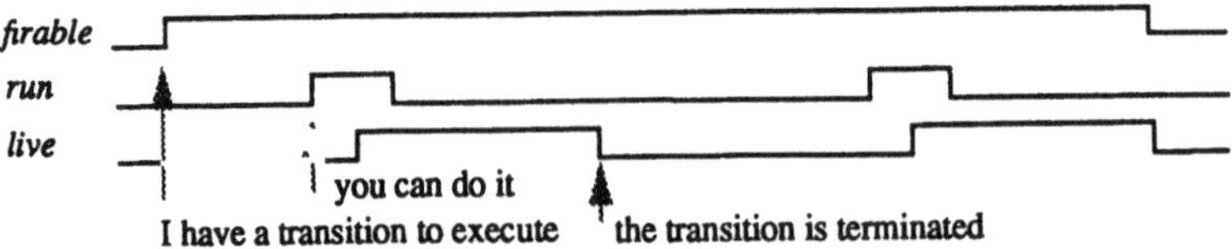

Figure 9: Waveforms Implementing the Parent/Child Synchronization

When there is no firable transition in a parent module, it can allow children to work:

- asynchronously (the parent has *systemactivity* or *activity* attribute, and there is no dynamism and no transition conditioned by when and priority clauses),
- sequentially (the parent has *systemactivity* or *activity* attribute, and the above condition is not satisfied),
- in parallel (the parent has *systemprocess* or *process* attribute).

If the corresponding child entities work asynchronously, then the *firable* signal can be omitted. If the child entities work in a sequential way, then a selection algorithm must be implemented to choose a child instance to work, e.g., round robin selection. If the child entities work in parallel, then one common *run* signal may be used for all of them. The parent entity must wait until all *live* signals (from its children) indicate the end of their actions.

In most of the designs, the hardware functional blocks work in parallel to obtain high performance. Usually, they stop only to wait on new data or to gain access to shared resources. In the Estelle computational model, a parent and child instances work sequentially to avoid competition to access shared (exported) variables. The sequential execution of tasks is closer to software implementation, where the task computation is not be performed simultaneously, because of processor time sharing. It is not recommended to realize hardware blocks, which work at mutually exclusive time periods. Hence, the Estelle modules which are planned to be realized in hardware should be allowed to work simultaneously, e.g., only the leaves of the module hierarchy tree should be allowed to have a transition part.

The VHDL representation of a message can only be a value carried by a signal. An Estelle connection link can carry two messages in opposite directions at the same moment. A VHDL signal cannot carry two values at the same time, hence two signals should be declared to represent a connection link. An Estelle channel declaration specifies all possible messages, which can be sent in both directions through a connection link. A pair of VHDL type declarations can correspond to an Estelle channel declaration. Each of these types enumerates values, which are allowed to be sent in one direction.

A FIFO queue, which keeps the incoming messages, should be declared as an entity as there is no equivalent object in VHDL. Figure 10 shows possible structures of queue entities.

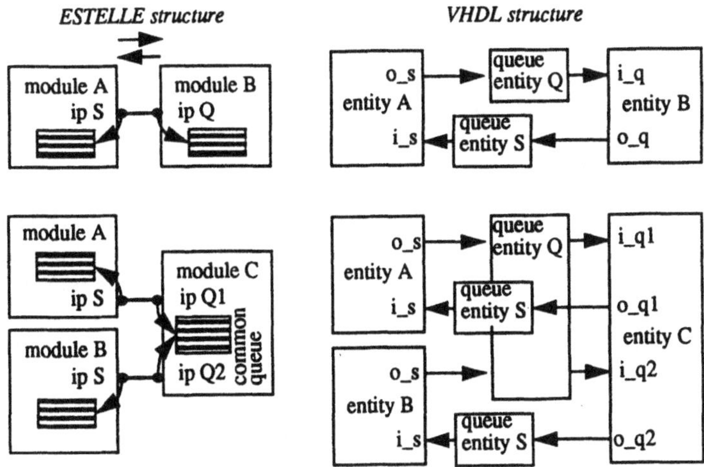

Figure 10: An Estelle Interface Translation to VHDL Entity Connections

Since there is no dynamism in VHDL, as a hardware structure is static, a dynamic behavior of Estelle modules may correspond to dynamic hardware reconfiguration. A reset signal may be used to freeze an entity activeness, which may represent termination of a corresponding module. Consequently, when the reset signal changes its value to false, it enables an entity activeness starting from the initial point - that represents the creation of a module. This way of modeling implies that the number of entities is constant and equal to the maximal number of possibly existing modules.

The dynamism of Estelle linking can be modeled by connection switching between entities. Different connection switching implementations are possible: e.g., three state buses, multiplexers and demultiplexers, or analog switches. Sophisticated algorithms are needed to carry on entity allocation. The allocation corresponds to a creation of a module instance during a protocol processing.

4.3. ESTELLE TO VHDL TRANSLATION MODEL

This chapter describes the translation model that is implemented in our e2v (Estelle to VHDL) translator. The non-obvious design decisions are discussed, as well as translation limitations caused by semantic and syntax differences.

The translation model presentation consists of two parts. The first part deals with architectural elements of an Estelle specification, i.e., channels, modules, interaction points, and queues. The second part explains synchronization and time model for a generated set of entities. This part deals with behavioral elements of an Estelle specification, i.e., transition clauses. It is not possible to give, in this short paper, the detailed mapping of all Estelle constructors to VHDL text. An exhaustive description of the translation model and the e2v translator manual is given in [7].

4.3.1. Architectural Part of Specification

4.3.1.1. Channels

An Estelle channel declaration (which defines two sets of messages, each related to a channel role) is translated to a pair of VHDL type declarations. Listing 1 shows an example of an Estelle channel declaration and generated VHDL data type definitions. These types correspond to the channel roles. If a channel role does not list any messages, then the corresponding VHDL type is not generated. Such a data type must carry a message name and all parameters which belong to each message of the role under translation. When there is no parameter, the data type may be an enumeration type (it defines the allowed message names). Otherwise, the data type must be a record type that assembles:

- the field for a message name - an enumeration type,
- the fields for parameters of each message - record types.

There are no variant parts at a record type definition in VHDL, so for any message with parameters, a fixed field must be declared. The field type should be a record that assembles all parameters of the message. This superfluity of data representation will be suppressed during the later synthesis steps, when the complex data types will be converted to bit_vector types.

The channel declarations are inherited by child modules, and therefore the corresponding type declarations are placed in VHDL package constructors[1].

4.3.1.2. Modules and Interaction Points

An Estelle module definition is composed of a module header declaration and of a module body declaration. Several body declarations may be given for one header declaration. A header is translated to an entity declaration, and a body is translated to an architecture declaration. Several architecture declarations may be given for one entity declaration.

An Estelle *init* statement creates a module instance, determining the module header and body constructors which describe this instance. A corresponding entity instance is conceived by VHDL *component instantiation* statement. A *configuration* statement binds the component instance with an entity and an architecture constructors. Listing 2 shows these instantiation statements.

A module header contains interaction points declaration and exported variables declaration. The two kinds of communication elements are translated to a signal port list declaration. A module header may contain a parameter list. These parameters serve to

---

[1] There is no heredity of data types in VHDL, but instead a package constructor exists which contains declarations. These declarations are intented to be visible in some entities.

differentiate the behavior of module instances. The right equivalent of the parameter list is the VHDL generic list, which fulfils the same purpose. Listing 3 shows an example of a module translation.

An Estelle body declaration may contain internal interaction points declaration. The internal interaction points, like the external (listed in a module header), are translated to VHDL signals. A VHDL architecture corresponding to the Estelle body contains these signals declarations, as Listing 3 shows.

4.3.1.3. FIFO Queues

Estelle FIFO queues are created together with module instances. Declarations of external and internal interaction points determine the queues and their kind (common or individual). Estelle queues are translated to VHDL entities. Some limitations must be introduced and some design choices must be done for this translation.

The abstract notion of an unbounded FIFO may model a receiver device of any complexity, which is connected with a sender by any sort of physical link (e.g., optical, electric, electromagnetic). It is not possible to describe in VHDL such an abstract object. The translation target is to preserve the behavior of the system under design and to generate a VHDL text, which will be useful in the subsequent synthesis steps. It is obvious that the queue entities will be modified during subsequent synthesis steps to fit realization needs.

A generated queue entity has a fixed size, given as a generic parameter - the lack of unbounded data types in VHDL causes this limitation. The next translation problem arises due to the lack of variable data types in VHDL (messages stored in a queue, which are usually data of different types). All VHDL signals have fixed data types, and thus also the signals from the entity port list, which correspond to interaction links. There are two solutions to this problem:

- A separate queue entity declaration is generated for each queue that stores different type messages (declared for different channel roles).
- One queue entity declaration is given with a uniform signal data type, e.g., bit_vector. A pair of functions must be given to perform a conversion to and from the uniform data type for each data type corresponding to an interaction point role.

The first solution gives a shorter VHDL text; moreover, some applications may need different queue realizations for different functional blocks. We adopt the first solution, and we propose two entity class for queue representation: individual queue entity class and common queue entity class. Entities belonging to the same class differ only by the types of input-output signals and by types of stored data.

To put (to get) a message on (from) a queue, control signals are needed. Some realizations of control signals may be implemented. The simplest realization may have only one put and one get signal, and a separate signal to carry the mark (name) of a message. To ensure the correct data write (read) to (from) a queue, two solutions are possible:

- the put and get signals must satisfy setup and hold times in relation to data signal,

• two acknowledge signals must be introduced to realize a handshake mechanism.

We chose the handshake mechanism as the more reliable solution. Figure 11 shows the appropriated waveforms.

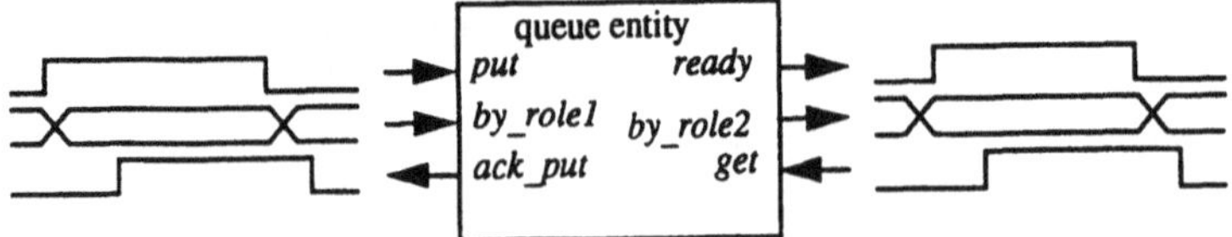

Figure 11: Waveforms of Queue Control Signals

Two or one queue entity is generated during translation of an Estelle channel declaration. The entities correspond to the channel roles.

The particular realization of an interaction points connection may be completely different than that proposed above, e.g., a serial, bidirectional data link may be used with a pair of receivers/transmitters and serial/parallel converters. In such a case, the designer must substitute the queue entities by a set of more detailed entities. The real data path type is bit_vector, and it may be unidirectional or bidirectional, e.g., three-state. The modification of entity ports declaration must be done by a designer manually during the later synthesis step. However, when the proposed queue model is satisfactory, it may be used to direct synthesis. The queue entity proposed above fulfills the Estelle semantics. It may be used to generate VHDL simulation patterns for verification of later design shape.

4.3.2. Behavioral Part of Specification

An Estelle specification describes a set of modules. A module may be implemented in hardware as a synchronous automaton or as an asynchronous automaton.

The synchronous realizations are dominant because of designing process simplicity (shorter designing time and smaller probability of design errors). The only advantage of asynchronous realization is the possibility of gaining a better circuit performance. Taking into account the clock frequencies that are applied to modern digital circuits, asynchronous realization is no longer used much, especially in the case of a complex design.

A module with a non empty transition part is an extended finite state automaton, and a transition execution may involve computation of any complicated algorithm. A corresponding functional block should be a processor with an execution part specialized for performing a given algorithm. Its control part should perform:

- the selection of an action to execute,
- the data passing control inside the execution part,
- the synchronization with queue entities, and
- the synchronization with parent and child entities to satisfy the parent-child priority.

It is obvious that such a complicated functional block should be synchronous, i.e., all internal transactions are performed synchronously to a clock signal. Moreover, synchronous designs are more reliable and easy to debug. Introducing a clock to VHDL description of a module simplifies the description. Without a clock signal, it is very difficult to write a correct and readable VHDL process that has wait loops for external events (e.g., *ack_put* signal from queue entities, look at Figure 11) and is sensitive to other signals (e.g., an internal state signal, *ready* signals from queue entities, and *run* signal from parent entity). Moreover, the clock signal is useful to implement an Estelle delay clause.

An Estelle transition description is composed of a set of transition clauses and a compound statement. Intermediate steps of a clock signal are needed to go through internal states of the implementation, which computes the compound statement given for an Estelle transition. Figure 12 shows an example of internal states which implement Estelle transitions. The internal states are irrelevant to the states defined in an Estelle specification, they depend on the implementation of the transition compound statements. The complexity of a transition compound statement determines the time (number of clock steps) needed to execute the corresponding action.

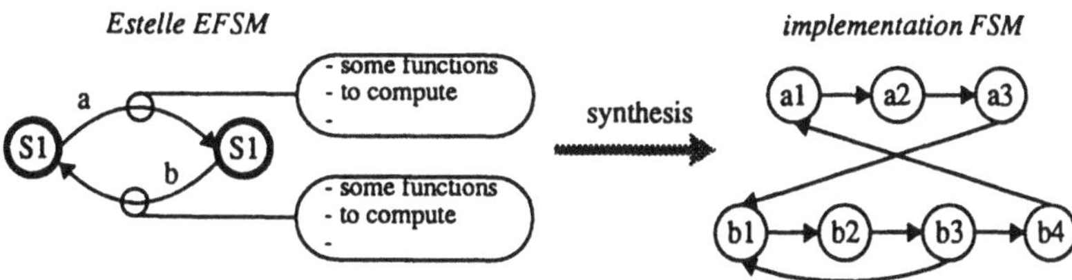

Figure 12: Example of Implementation FSM that Corresponds to an Estelle EFSM

We assume that each system module has its own clock generator. The clock frequency is determined by a generic value: Tw, i.e., clock pulse width. The clock pulse width is defined in time unit 1000 times smaller than the time unit given by Estelle *timescale* statement. This arbitrary selected clock pulse width can be refined during synthesis. Each computation step which corresponds to an Estelle transition selection starts synchronously to the clock signal. An action corresponding to an Estelle transition execution goes on one or more clock periods.

The initialization and the transition parts of a module describe the module behavior. The initialization part is a set of sequentially executed statements, which are executed once at the creation instant. Firable transitions are executed sequentially in a nondeterministic order. There is no parallelism inside one module, and therefore the transitions are not translated to concurrently working VHDL processes. The Estelle initialization and transition parts are represented in VHDL as one *process* statement, as Listing 4 shows.

The Estelle semantics allows several transitions to be firable in a given time. In VHDL a transition is represented by an action. Different algorithms can be implemented to select an action to perform, between these that are ready. Our translator generates a simple algorithm, which selects the first ready action from the action list, as Listing 5 shows. A designer may change this algorithm to comply with the different implementation needs. The input to the selection function is the *tr_cond* vector of boolean values. Each element of this vector corresponds to one defined transition. An element has a true value if the corresponding transition is firable. The selection function returns the number of an action to execute.

The advantage of this algorithm is the implementation simplicity. This algorithm introduces a linear priority of actions. The first action on the list has the highest priority, the last action has the lowest. In some cases, this facility cannot be accepted. For example: assume a module with two always ready transitions. In the Estelle computation model, the two will be executed, due to indeterministic choice. In our implementation, only one action (that corresponding to the first transition) will be executed.

All local variables of a module are translated into the *process* statement declarative part. A state signal is generated for any module containing a state declaration. VHDL variables cannot be observed during simulation, so for this reason we chose a signal, not a variable, for module state representation.

4.3.2.1. Transition Clauses

An Estelle transition description is composed of a set of transition clauses and of a compound statement. The clauses condition the transition execution, i.e., the compound statement evaluation. There are six transition clauses in Estelle:

- *from* clause,
- *when* clause,
- *provided* clause,
- *delay* clause,
- *priority* clause.

For any transition a boolean expression is generated. The boolean expression is generated from the three clauses: *from*, *when*, and *provided*. The boolean expression is a conjunction of sub-expressions, which are generated individually for each present clause. Listing 6 shows some examples of an Estelle transition translation.

A *from* clause leads to verification of a value of state signal. A *when* clause leads to the following results:

- boolean expression, which test the presence of a valid message on a queue entity output and the name of the message,
- the *in_synchro* procedure call.

The *in_synchro* procedure gets a message from a queue entity. The *in_synchro* call follows the sequence of statements, which correspond to a transition compound statement.

A *provided* clause may contain a boolean expression or the *otherwise* keyword. The expression is directly translated to a VHDL boolean expression. The *otherwise* keyword is translated to a logical negation of the conjunction of all other *provided* expressions.

A *delay* clause may specify one or two parameters. The first parameter states: when a transition can (but must not) be executed. The second parameter states: when the transition becomes selected like other transitions (not delayed). A *delay* clause with one parameter is semantically equivalent to the clause with two identical parameters. Two different parameters serve to introducing nondeterminism in the selection of a delayed transition. This nondeterminism must be suppressed during translation to VHDL. Translation of a *delay* clause precipitates a timer declaration (a VHDL concurrent procedure call) and four signal declarations to control the timer.

A *priority* clause influences the transition selection. A priority constant parameter serves to partially order transitions. The implemented selection algorithm should consider this constant. The algorithm generated by our translator orders all actions in a priority list. The priority parameters are taken into account during creation of this list.

4.4. CONCLUSIONS

The feasibility of Estelle to VHDL translation has been analyzed, and the possible representations of Estelle items have been presented in this paper. The mapping examples of Estelle syntax elements to the VHDL description have been given. The conclusions of this analysis are as follows:

- It is possible to translate all Estelle semantics' items into VHDL, but any nondeterminism which can occur in an Estelle specification must be suppressed. The generated VHDL design works in a deterministic fashion:
  - Nondeterministic choice with respect to a selection of firable transitions is translated to a deterministic choice of ready actions.
  - Nondeterministic choice of a child module to work is translated to asynchronous parallel execution, when possible. Otherwise it is translated to deterministic sequential selection.

- Since there is no dynamism in VHDL, as a hardware structure is static, a dynamic behavior of Estelle modules may correspond to dynamic hardware reconfiguration.

- All Estelle modules intended for hardware implementation are translatable to VHDL.

The same system may be specified in Estelle in different ways. The way of specifying influences the complexity of generating hardware. There are some Estelle constructors that increase this complexity. A declaration of an exported variable which is modified by two modules causes more complex hardware implementation than a declaration of two exported variables, which are modified separately by respective modules. A common queue that is associated with two interaction points is more complex in hardware implementation than two individual queues. Hierarchy of active modules causes parent-child synchronization that slows down computation speed and increases implementation complexity.

Some VHDL constructs do not synthesize well. Different synthesis platforms have different restrictions to VHDL. There are some restrictions common to most of these synthesis tools:

- no block and no guarded block statements,
- no aliases,
- no records,
- no real-type,
- no recursive calling of procedures and functions,
- only one-dimensional arrays.

We took the two first restrictions into account during creation of the translation model. The three last restrictions within the above list can be followed during Estelle specification. It is obvious that a designer will have to do some work to adopt a generated VHDL text to be accepted by a synthesis tool. He should define conversion functions from composed data types (e.g., record types) to bit-vector types and expand or modify the queue entities to desired realizations of these functional blocks. He should introduce all details relevant with a concrete implementation.

A presented Estelle to VHDL translator enables a passage from protocol specification and analysis tools to hardware simulation and synthesis tools, as Figure 13 shows.

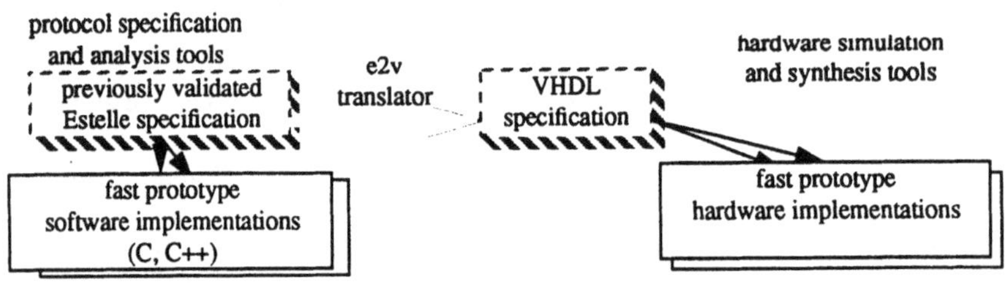

Figure 13: Link between Software and Hardware Designing

There are three advantages of this approach:

- The hardware and software implementations are based on the same global system specification.
- The generated VHDL text can be used to produce some VHDL simulation scenarios and waveform results, which can be used to design validation at the later synthesis steps.
- The VHDL simulation tools can be used to detailed analysis of time dependences and performance of a target system.

The Estelle language is more efficient than the VHDL language in describing a distributed system and analyzing a communication protocol in the beginning phase of designing. The Estelle specification is more compact than the VHDL specification (more than 40%). VHDL tools, in contrast to Estelle tools, are well suited to analyze time dependencies. For example, the Estelle simulation tools are not able to take into account the delay times of a message transfer. We allow a user to indicate time parameters inside an Estelle specification using so-called qualifying comments. These parameters are conveyed into generated VHDL text. Consequently, our approach can be useful not only for hardware synthesis purposes.

During synthesis steps, signal types are changed to the bit_vector type. The simulation results of the same design in the different synthesis stages can be easily compared by building a signal converter. Figure 14 shows the known idea of such a signal converter. An entity which is transformed by synthesis steps must be encapsulated with a signal converter to obtain compatible signals.

The e2v translator works in the EDT (Estelle Development Tools) environment. The input to e2v is an Estelle Intermediate Form file. The Intermediate Form file is generated by the Estelle to C (ec) compiler [2] from an Estelle text file. The e2v output VHDL files were tested (compiled and simulated) in the VHDL environment from Vantage Analysis Systems, Inc.

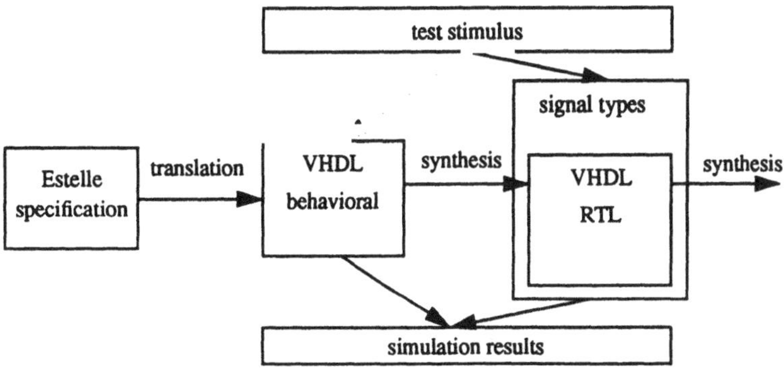

Figure 14: Signal Converter Entity Placement

REFERENCES

[1] S. Budkowski, P.Dembinski, "*An Introduction to ESTELLE: A Specification Language for Distributed Systems*" Computer Networks and ISDN Systems, vol.14, no.1, 1987.

[2] S. Budkowski, "*Estelle development toolset (EDT)*" Computer Networks and ISDN Systems 25, pp.63-82, 1992.

[3] W.Glunz, T.Kruse, T.RVssel, D.Monjau, "*Integrating SDL and VHDL for System-Level Hardware Designing*" pp.187-204, IFIP Transactions "Computer Hardware Description Languages," D.Agnew, L.Claesen, R.Camposano (Editors), Elsevier Science Publishers B.V., 1993.

[4] "*1076-1987 Standard VHDL Language Reference Manual*", IEEE Press, The IEEE computer society press 1991.

[5] International Organization for Standardization, IS 9074, "Information Processing Systems - Open System Interconnection - *ESTELLE: A Formal Description Technique Based on an Extended State Transition Model*" 1989.

[6] C.Delgado Kloos, T.de Miguel Moro, T.Robles Valladares, G.Rabay Filho, A.Marmn Lspez, "*VHDL Generation from a Timed Extension of the Formal Description Technique LOTOS within the FORMAT project*" Microprocessing and Microprogramming 38, pp.589-596, North-Holland, 1993.

[7] J.Wytrebowicz, "*VHDL Generation from Estelle*" Research report No 941103, Institut National des Télécommunications, Evry, October 1994.

APPENDIX

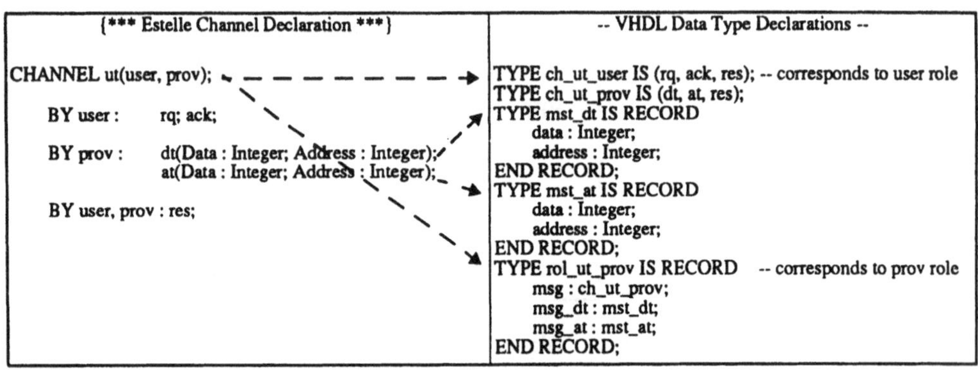

```
{*** Estelle Channel Declaration ***}

CHANNEL ut(user, prov);
    BY user :    rq; ack;
    BY prov :    dt(Data : Integer; Address : Integer);
                 at(Data : Integer; Address : Integer);
    BY user, prov : res;
```

```
-- VHDL Data Type Declarations --

TYPE ch_ut_user IS (rq, ack, res); -- corresponds to user role
TYPE ch_ut_prov IS (dt, at, res);
TYPE mst_dt IS RECORD
    data : Integer;
    address : Integer;
END RECORD;
TYPE mst_at IS RECORD
    data : Integer;
    address : Integer;
END RECORD;
TYPE rol_ut_prov IS RECORD    -- corresponds to prov role
    msg : ch_ut_prov;
    msg_dt : mst_dt;
    msg_at : mst_at;
END RECORD;
```

Listing 1: Example of Channel Translation

```
{*** Estelle Module Instantiation ***}
MODVAR
    rcvr : receiver;
{* text irrelevant to this example *}

INIT rcvr WITH receiver_body(init_rcvr_to_active,
    listen_confirm);
```

```
-- VHDL Component Instantiation --
rcvr : receiver
    GENERIC MAP ( init_rcvr_to_active, listen_confirm)
    -- text irrelevant to this example
                -- VHDL Configuration --
FOR rcvr : receiver
    USE ENTITY work.receiver(receiver_body);
END FOR;
```

Listing 2: Example of a Module Instantiation Translation

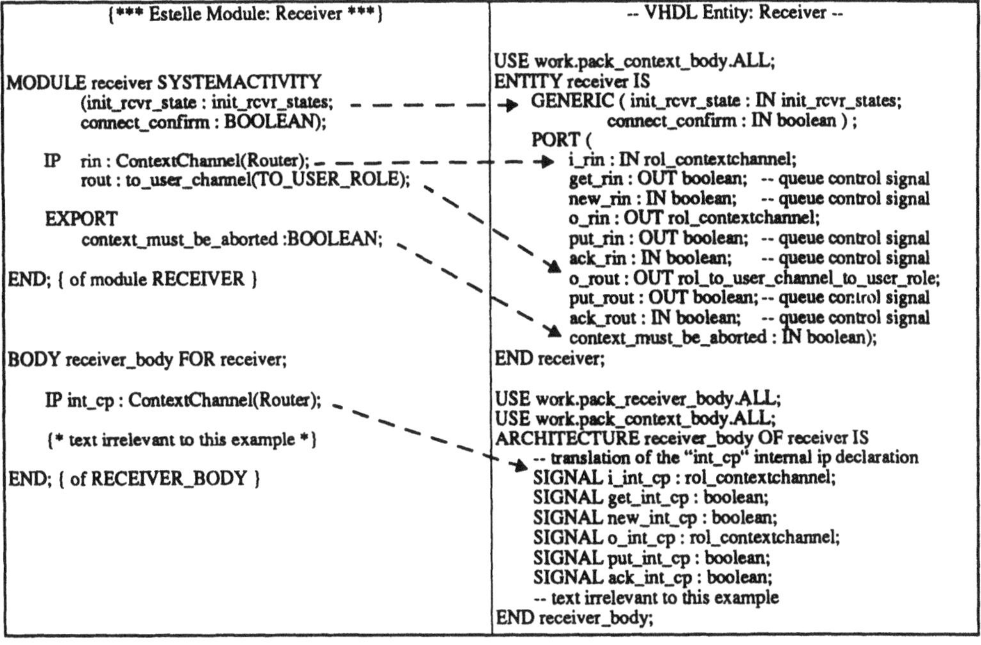

```
{*** Estelle Module: Receiver ***}

MODULE receiver SYSTEMACTIVITY
       (init_rcvr_state : init_rcvr_states;
       connect_confirm : BOOLEAN);

    IP  rin : ContextChannel(Router);
        rout : to_user_channel(TO_USER_ROLE);

    EXPORT
        context_must_be_aborted :BOOLEAN;

END; { of module RECEIVER }


BODY receiver_body FOR receiver;

    IP int_cp : ContextChannel(Router);

    {* text irrelevant to this example *}

END; { of RECEIVER_BODY }
```

```
-- VHDL Entity: Receiver --

USE work.pack_context_body.ALL;
ENTITY receiver IS
    GENERIC ( init_rcvr_state : IN init_rcvr_states;
              connect_confirm : IN boolean ) ;
    PORT (
        i_rin : IN rol_contextchannel;
        get_rin : OUT boolean;   -- queue control signal
        new_rin : IN boolean;    -- queue control signal
        o_rin : OUT rol_contextchannel;
        put_rin : OUT boolean;   -- queue control signal
        ack_rin : IN boolean;    -- queue control signal
        o_rout : OUT rol_to_user_channel_to_user_role;
        put_rout : OUT boolean; -- queue control signal
        ack_rout : IN boolean;   -- queue control signal
        context_must_be_aborted : IN boolean);
END receiver;

USE work.pack_receiver_body.ALL;
USE work.pack_context_body.ALL;
ARCHITECTURE receiver_body OF receiver IS
    -- translation of the "int_cp" internal ip declaration
    SIGNAL i_int_cp : rol_contextchannel;
    SIGNAL get_int_cp : boolean;
    SIGNAL new_int_cp : boolean;
    SIGNAL o_int_cp : rol_contextchannel;
    SIGNAL put_int_cp : boolean;
    SIGNAL ack_int_cp : boolean;
    -- text irrelevant to this example
END receiver_body;
```

Listing 3: Example of a Module Translation

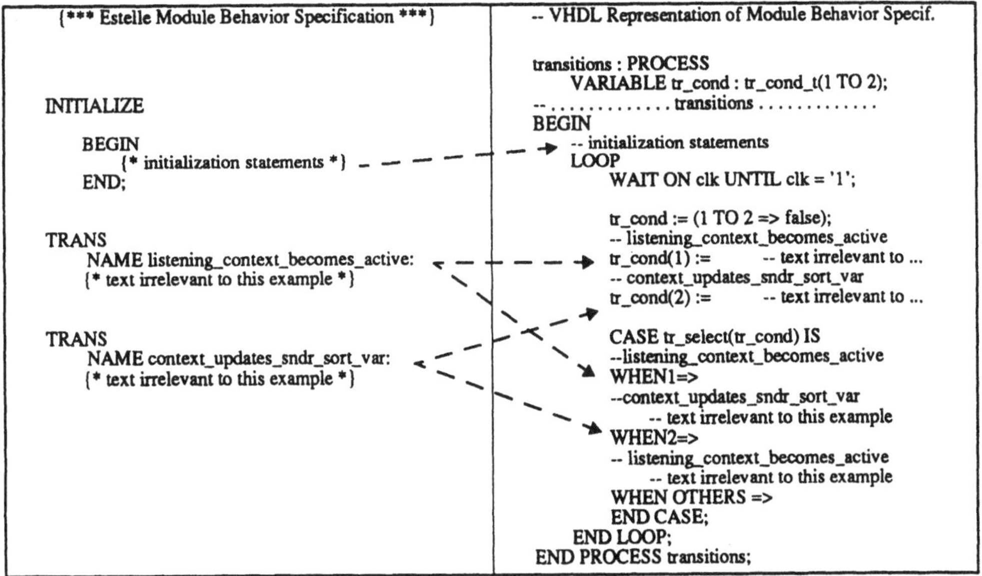

```
{*** Estelle Module Behavior Specification ***}

    INITIALIZE

        BEGIN
            {* initialization statements *}
        END;

    TRANS
        NAME listening_context_becomes_active:
        {* text irrelevant to this example *}

    TRANS
        NAME context_updates_sndr_sort_var:
        {* text irrelevant to this example *}
```

```
-- VHDL Representation of Module Behavior Specif.

transitions : PROCESS
    VARIABLE tr_cond : tr_cond_t(1 TO 2);
-- . . . . . . . . . . . . transitions . . . . . . . . . . . .
BEGIN
    -- initialization statements
    LOOP
        WAIT ON clk UNTIL clk = '1';

        tr_cond := (1 TO 2 => false);
        -- listening_context_becomes_active
        tr_cond(1) :=        -- text irrelevant to ...
        -- context_updates_sndr_sort_var
        tr_cond(2) :=        -- text irrelevant to ...

        CASE tr_select(tr_cond) IS
        --listening_context_becomes_active
        WHEN1=>
        --context_updates_sndr_sort_var
            -- text irrelevant to this example
        WHEN2=>
        -- listening_context_becomes_active
            -- text irrelevant to this example
        WHEN OTHERS =>
        END CASE;
    END LOOP;
END PROCESS transitions;
```

Listing 4: Example of Initialization and Transition Parts Translation

```
TYPE tr_cond_t IS ARRAY (integer RANGE <>) OF boolean;
FUNCTION tr_select (tr_cond : IN tr_cond_t          -- transition state vector
                   ) RETURN integer IS              -- selected transition number
    BEGIN
        FOR i IN tr_cond'RANGE LOOP
            IF (tr_cond(i)) THEN RETURN i;
            END IF;
        END LOOP;
        RETURN 0;
    END tr_select;
```

Listing 5: Transition Selection Function

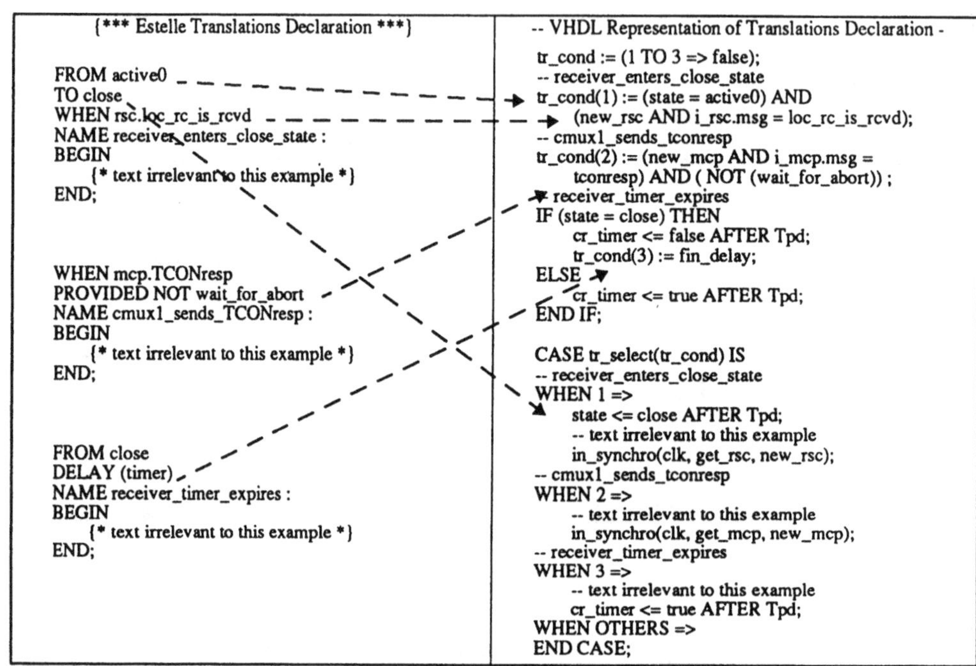

```
{*** Estelle Translations Declaration ***}

FROM active0
TO close
WHEN rsc.loc_rc_is_rcvd
NAME receiver_enters_close_state :
BEGIN
    {* text irrelevant to this example *}
END;

WHEN mcp.TCONresp
PROVIDED NOT wait_for_abort
NAME cmux1_sends_TCONresp :
BEGIN
    {* text irrelevant to this example *}
END;

FROM close
DELAY (timer)
NAME receiver_timer_expires :
BEGIN
    {* text irrelevant to this example *}
END;
```

```
-- VHDL Representation of Translations Declaration -
tr_cond := (1 TO 3 => false);
-- receiver_enters_close_state
tr_cond(1) := (state = active0) AND
    (new_rsc AND i_rsc.msg = loc_rc_is_rcvd);
-- cmux1_sends_tconresp
tr_cond(2) := (new_mcp AND i_mcp.msg =
    tconresp) AND ( NOT (wait_for_abort)) ;
-- receiver_timer_expires
IF (state = close) THEN
    cr_timer <= false AFTER Tpd;
    tr_cond(3) := fin_delay;
ELSE
    cr_timer <= true AFTER Tpd;
END IF;

CASE tr_select(tr_cond) IS
-- receiver_enters_close_state
WHEN 1 =>
    state <= close AFTER Tpd;
    -- text irrelevant to this example
    in_synchro(clk, get_rsc, new_rsc);
-- cmux1_sends_tconresp
WHEN 2 =>
    -- text irrelevant to this example
    in_synchro(clk, get_mcp, new_mcp);
-- receiver_timer_expires
WHEN 3 =>
    -- text irrelevant to this example
    cr_timer <= true AFTER Tpd;
WHEN OTHERS =>
END CASE;
```

Listing 6: Example of Transition Clauses Translation

5

AN ALGORITHM FOR THE TRANSLATION OF SDL INTO SYNTHESIZABLE VHDL

R.J.O. Figueiredo and I.S. Bonatti

Universidade Estadual de Campinas, UNICAMP, Faculdade de Engenharia Elétrica, Departamento de Telemática, Campinas, Brazil.

ABSTRACT

The high-level behavioural description of digital systems allows the designer to focus on ***what*** *the circuit is supposed to do instead of* ***how*** *it will be implemented. This letter presents a means of using SDL as a front-end, high-level graphical description tool by translating a subset of this language into fully synthesizable VHDL.*

5.1. INTRODUCTION

The gap between specifications and implementation of digital circuits is responsible for a good deal of design effort spent in the mapping of a given specification into a real system. The development of behavioural Hardware Description Languages, especially VHDL [9] [10], and the availability of Electronic Design Automation (EDA) tools that make use of such languages have allowed the description of hardware at a higher level. Gate-level synthesis from a VHDL description is straightforward if the VHDL code respects the synthesis constraints.

The goal of this work is to place the hardware design closer to its specifications, without losing the implementation capability. This will be accomplished by using

CCITT's SDL (Specification and Description Language [1]) as a graphical description tool.

A proposal for using SDL for VHDL-based hardware design was first introduced in [4], and further refined in [7], covering simulation aspects, and in [3], covering synthesis. In [6] an intermediate language interfacing system-level (SDL) and implementation-level (VHDL) languages is presented, and a software/hardware co-design approach based on this language is proposed in [5]. Further work in this topic can be found in [11] and [8].

This article proposes the use of a subset of the SDL language, consisting of its most common elements (system, blocks, processes, states, signals, tasks and decisions) for the hardware design and the application of certain constraints to achieve efficient synthesis.

The first part of this article focuses on which elements of SDL are supported, how to represent them in VHDL and the constraints that are applied to this SDL subset to make the description efficiently synthesizable. The next part covers the translation algorithm itself. It is assumed that the reader is familiar with the basic syntax and semantics of SDL and VHDL.

5.2. THE SDL SUBSET

Some SDL features are not currently supported, such as services and procedures. However, the most common SDL elements are supported by the algorithm, and are well suited to the behavioural description of digital systems:

- ***System***: The system corresponds to the highest-level structure of the circuit being developed, and maps into a VHDL entity.
- ***Block*** and ***Block Substructure***: Blocks provide the structure of the design, and are also mapped into VHDL entities.
- ***Process***: Processes are supported, but restricted to only one instance per process; dynamic creation of processes is not allowed. Each SDL process maps into two entities: a VHDL entity containing a process representing the behaviour of the SDL process, and another entity containing a synthesizable representation of the SDL signal-based communication scheme.
- ***State***: VHDL enumerated types are used to represent states.
- *Decision*: Decisions are mapped into VHDL IF-THEN-ELSE branches.
- ***Task***: Assignment tasks are supported, as well as logic and simple arithmetic operations (addition and subtraction) on the right side of the assignment. They are mapped into VHDL assignments and logic/arithmetic operations.
- ***Signal***: The interchange of signals among processes via INPUT/OUTPUT primitives is allowed. The infinite-length queue abstraction of SDL is not supported; finite queues are used instead. Continuous signals and enabling conditions are also supported.
- ***Channel, Route***: Communication paths for signal exchanging are mapped into netlist connections among the VHDL entities representing SDL blocks and processes.

- ***Variable***: Mapped into VHDL variables by default. The designer can specify that SDL variables be mapped into VHDL signals, in order to obtain better simulation results. REVEALED variables are also supported.
- ***Abstract types***: Supported types are: range-constrained integers, booleans and uni-dimensional arrays of boolean data.

5.3. SYNTHESIS CONSTRAINTS

Many SDL features would lead to a very costly implementation if they were to be represented in hardware. For example, the dynamic creation of processes, floating-point arithmetic and infinite-length signal queues would require a large number of components and are cost-prohibitive; they are not supported.

5.4. THE MAPPING ALGORITHM

The hardware implementation of the SDL system is synchronous to a global clock that triggers every transition inside processes and every SDL signal sent or received. Synchronous machines are less sensitive to timing errors than asynchronous ones, and are more easily represented in synthesizable VHDL.

With this synchronization scheme every transition body in an SDL process takes one period of the master clock to be processed, and every signal sent from one process to another also takes a clock period to arrive in its destination.

The master clock is automatically allocated by the translation algorithm, as well as a master asynchronous reset that initializes all variables and starting states of processes. Both VHDL signals, master clock and reset, are input ports of the highest-level VHDL entity and form the sensitive list for every VHDL process of the system.

5.5. STRUCTURE

The SDL system generates a highest-level VHDL entity that instantiates all of its blocks as VHDL components and has input/output ports according to the signals sent to or received from the environment.

An SDL block generates a VHDL entity that instantiates all of its processes as VHDL components and has input/output ports according to the signals sent to or received from its channels. A block substructure also generates a VHDL entity instantiating its blocks.

An SDL process generates two separate VHDL entities. The first entity holds the behaviour of the corresponding SDL process. The second one implements the communication protocol associated with this process. There are currently three types of protocols available to the designer, described later in this text.

5.6. DATA DEFINITION

Types can be derived from the basic set (boolean and integer) by using either the SYNTYPE (for constrained integers) or the NEWTYPE construct (only for the definition of arrays); both map into VHDL types. SDL synonyms can also be defined, and are mapped into VHDL constants.

Sorts defined in an SDL system, block or block substructure are placed in VHDL packages, one for each scope, so that they can be further referenced by the lower-level blocks and processes. Sorts defined in an SDL process are defined in the corresponding VHDL entity modelling the process behaviour.

5.7. PROCESS COMMUNICATION

Each SDL process has one corresponding VHDL process modelling its signal-based communication. This process provides an interface between the incoming signals and the behaviour of the process. There are currently three kinds of communication protocols described in VHDL available to model the signal exchanging scheme of SDL:

1 *No queue*: There is only one slot reserved for *all* incoming SDL signals. If more than one signal arrives at the same time (ie, at the same master clock period), only one will be held. Priorities must be assigned to incoming SDL signals to solve this conflict.
2 *Single-position queue*: There is one slot reserved for *each* incoming SDL signal. The most recently received signal occupies this slot. If more than one signal is present, they will be consumed according to their order of priority.
3 *Finite queue*: There are *N* slots reserved for all incoming SDL signals. The queue follows the first-in-first-out (FIFO) policy.

The designer assigns one of these protocols to each SDL process keeping in mind implementation costs and the communication characteristics of each process. Protocol **1** is very small in terms of hardware, requiring just one D flip-flop and an XOR gate for each signal. The hardware complexity increases in protocols **2** and **3**.

The interface between protocol and process behaviour is the same for any protocol, allowing more protocols to be defined or further optimization of their VHDL implementation. Moreover, the behaviour-modelling VHDL process does not have to know which protocol is allocated for any other process, since the interface does not change.

The action of sending an SDL signal via the OUTPUT primitive is performed with the following actions (in the VHDL process modelling the SDL process behaviour):

- assignment of the values of each parameter, if any, to the respective VHDL output ports;
- logic inversion of the VHDL output port that represents the action of sending.

The VHDL process modelling the protocol of the receiving SDL process notices this request by sensing the logic inversion in one of its inputs, and notifies the arriving of the signal by performing the following actions:

- holding a positive logic pulse in the VHDL output port that represents the action of receiving, for one period of the master clock;
- keeping the values of the parameters available during this clock period.

A transition is initiated when the receiving VHDL process senses the presence of this positive pulse in the VHDL input port written by the protocol. This test is accomplished through the use of an IF-ELSE-ELSIF structure, ordered according to the priorities assigned by the designer to each incoming signal.

5.8. PROCESS BEHAVIOUR

Each SDL process has one corresponding VHDL process modelling its behaviour. The sensitive list of this process is solely the master clock and the master reset automatically created by the algorithm.

The state of a process is stored in a VHDL signal local to the entity. State transitions are modelled with a CASE/WHEN structure. Each possible state has a corresponding entry in a WHEN clause. The body of each WHEN clause has an IF-ELSIF-THEN construct that tests the presence of an incoming signal. The body of each IF-ELSIF construct has the transition body (actions) for the corresponding incoming signal.

Assignment tasks are mapped into sequential VHDL assignments. Decisions are mapped into IF-ELSE-THEN constructs. Outputs are mapped into the procedures stated in the subsection *Process Communication.*

5.9. VARIABLE DECLARATION

Variables declared in an SDL process are declared, by default, as variables in the behaviour-modelling VHDL process. The designer has the ability to declare them as VHDL signals for simulation purposes, taking into account the differences between the behaviour of a signal and a variable in VHDL.

Variables declared REVEALED are made visible to the other processes in the same block by creating an output port with a copy of their value. Entities that view revealed variables have input ports that are read when a VIEW access is made.

Since the algorithm does not allow dynamic process creation, the SDL type PId is not supported. Therefore, VIEW expressions must not contain a PId expression; the SDL-92 Z.100 [2] recommendation allows the resolution by name of the viewed variable.

5.10. EXAMPLE OF IMPLEMENTATION

The methodology of designing a digital circuit from a graphical SDL description was applied to the project of a Vending Machine. This example is inspired in the one used by Perry in [10] to illustrate the application of the Top-Down methodology with VHDL.

The vending machine accepts coins of four sorts (half dollar, quarter dollar, dime and nickel) and holds four kinds of products that can be purchased (pretzels, chips, cookies and doughnuts) each one having a different price.

The machine has a digital controller that keeps account of how much money has been entered, returns the change to the customer and keeps track of how many products are still left inside it.

The controller communicates with the customer via input buttons and output indicators (leds and decimal displays), and communicates with the rest of the machine via input sensors (sensing which sort of coin was inserted) and output mechanical devices (dispatching of the product purchased and of the change).

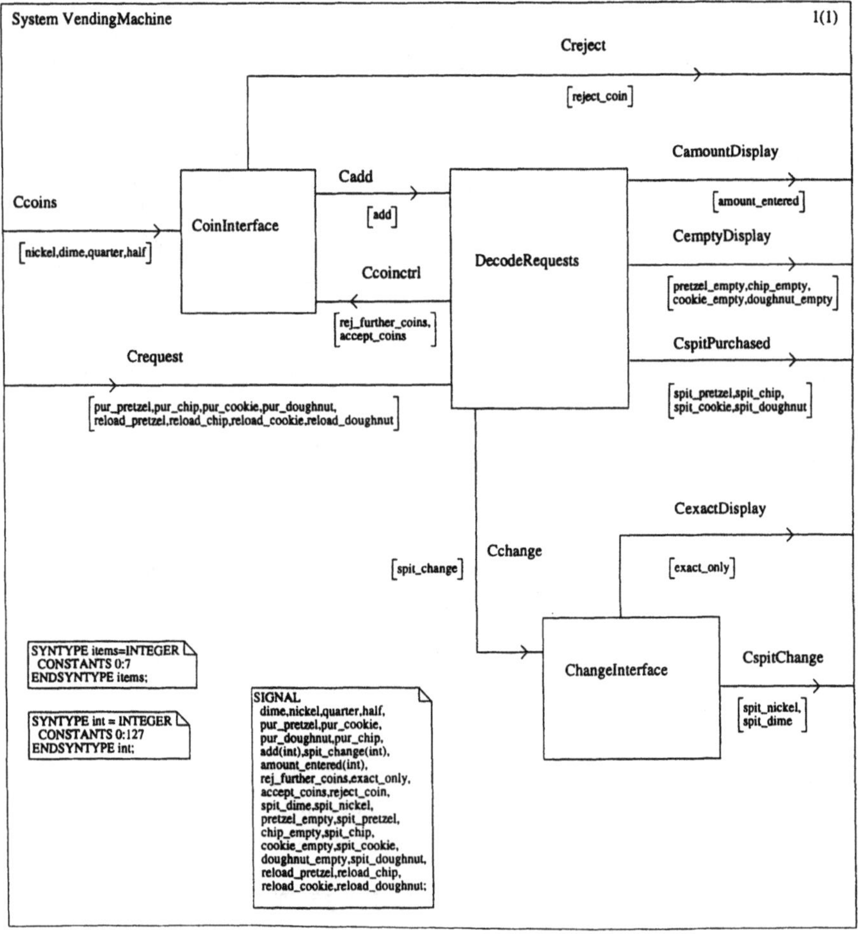

Figure 1: System VendingMachine

The system-level description of the controller (SDL system *VendingMachine*, Figure 1) shows this interface with the external environment, as well as the structure partition of the system in blocks and the signal channels linking them.

@Inside the SDL block *DecodeRequests* (Figure 2) are the processes that handle the amount of money entered (*AmountHandler*) and the purchase requests for each product type, as well as the signal routes that link them to the channel interconnecting blocks.

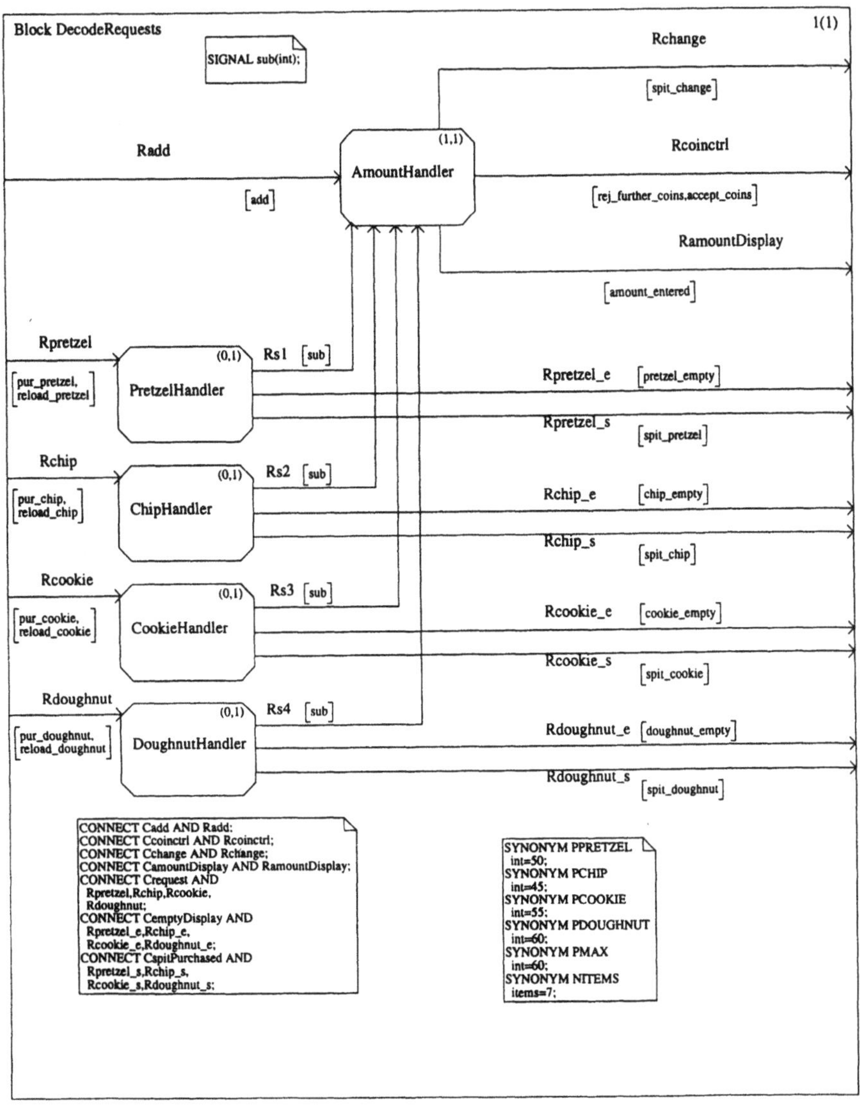

Figure 2: Block DecodeRequests

Inside the SDL process *ChipHandler* (Figure 3) is the finite state machine that models it. The process is usually waiting for a purchase request (state *pur_wait*). When the machine has no more chips, it warns the customer via an external indicator (signal *chip_empty*) and stays locked in the *empty* state until the machine maintenance service reloads the machine with chips (input *reload_chip*).

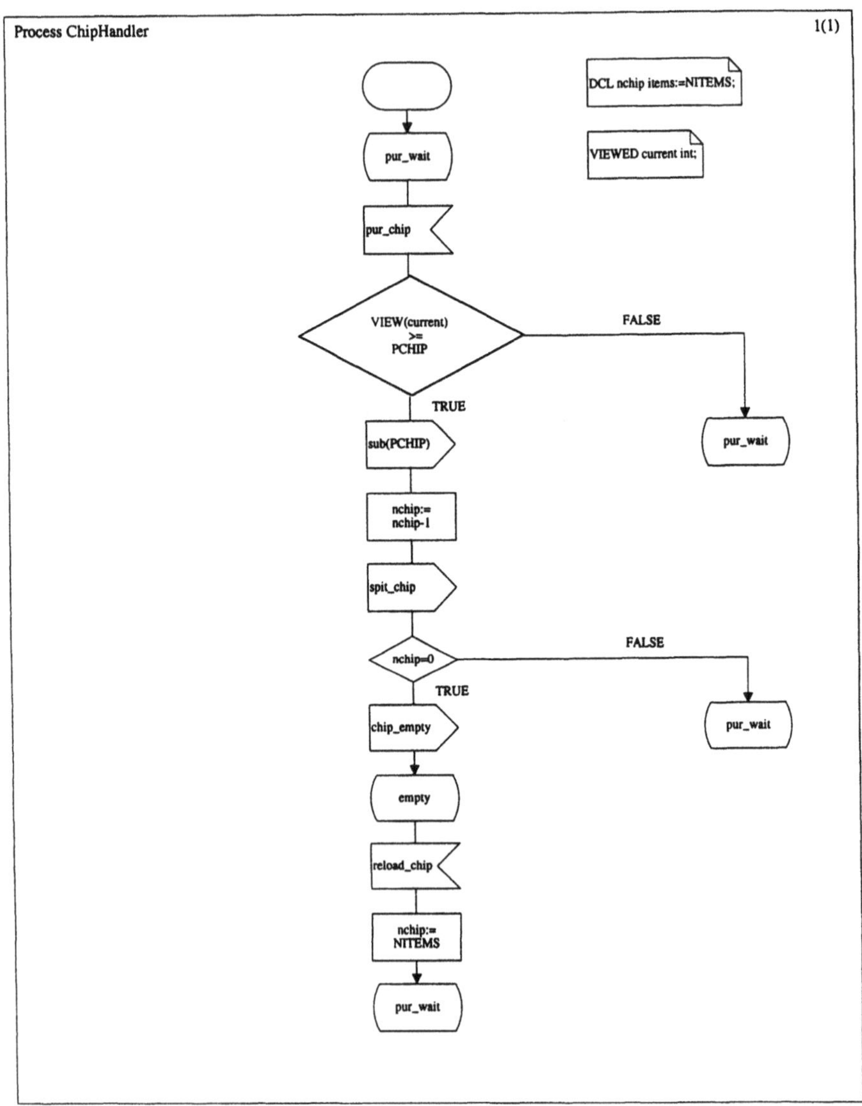

Figure 3: Process ChipHandler

The generated VHDL code for the behaviour-modelling process of *ChipHandler* is shown in Figure 4.

```
p_chiphandler : PROCESS(clock,reset)

    VARIABLE nchip : items;

  BEGIN

    IF (reset='1') THEN
      cstate_chiphandler <= pur_wait;
      nchip := NITEMS;

    ELSIF (clock'EVENT AND clock='1' AND
           clock'LAST_VALUE='0') THEN

      CASE cstate_chiphandler IS

        WHEN pur_wait =>
          IF (rec_pur_chip='1') THEN
            IF (revealed_current>=PCHIP) THEN
              send_param1_sub <= PCHIP;
              send_sub <= NOT(send_sub);
              nchip:=nchip-1;
              send_spit_chip <=
                NOT(send_spit_chip);
              IF (nchip=0) THEN
                send_chip_empty <=
                  NOT(send_chip_empty);
                cstate_chiphandler <= empty;
              END IF;
            END IF;
          END IF;

        WHEN empty =>
          IF (rec_reload_chip='1') THEN
            nchip := NITEMS;
            cstate_chiphandler <= pur_wait;
          END IF;

      END CASE;

    END IF;

  END PROCESS p_chiphandler;
```

Figure 4: VHDL for Process ChipHandler

The example above has been described in a graphical SDL editor (TeleLogic's *SDT*), converted into VHDL following the algorithm proposed in this article, simulated (behaviour), synthesised and re-simulated (RTL-level) within the Mentor Graphic's framework. The steps from the SDL description to the gate-level schematic representation are all automatic with the use of software tools. Figure 5 shows the synthesis result for the process ChipHandler.

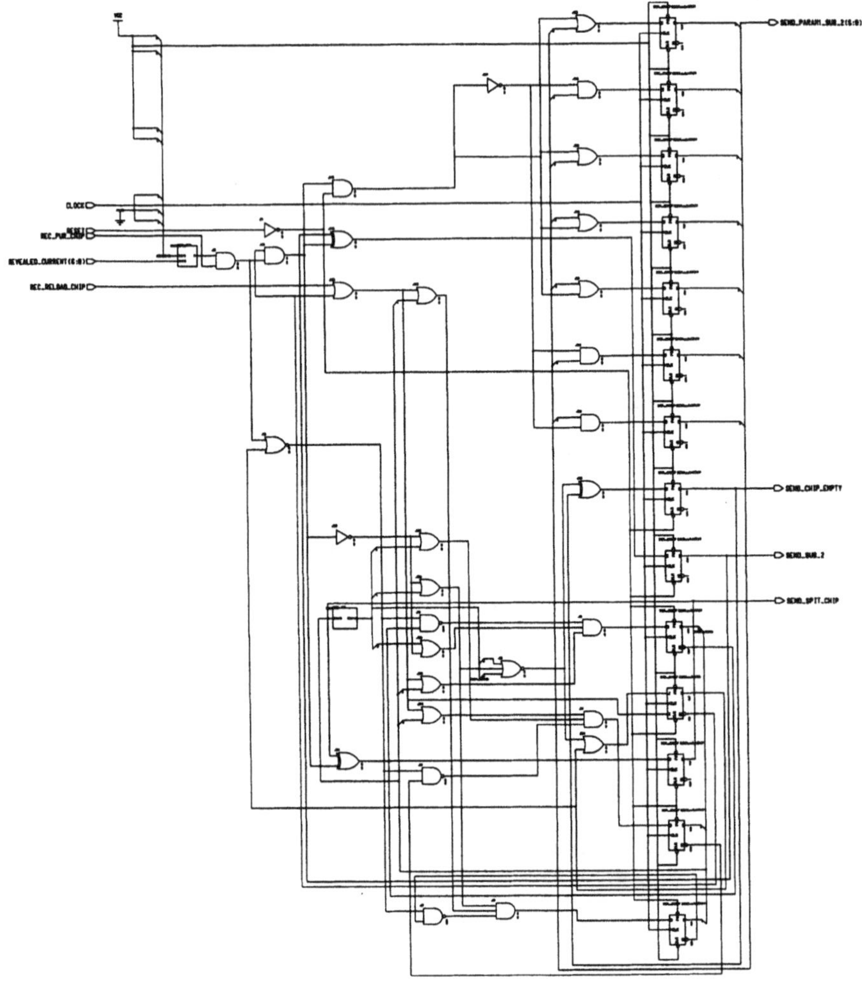

Figure 5: Process ChipHandler Synthesized

Table 1 shows, for this example, the gain in abstraction that this methodology introduces. The SDL-GR column shows the number of graphical symbols used to describe the system; SDL-PR and VHDL columns show the number of lines of code generated; the flip-flops and gate columns show statistics of the resulting synthesized circuit.

SDL-GR	SDL-PR	VHDL	flip-flops	gates
160	610	1470	140	570

Table 1: Gain in Abstraction

5.11. CONCLUSION

The specifications of digital systems at higher levels provides faster development cycles, technology-independent circuits and better documentation. The proposed algorithm allows hardware designers to use a subset of the high-level graphical language SDL as a front-end to their designs, by automatically generating fully synthesizable and syntatically correct VHDL. Further research will include aspects of hardware/software codesign and cover more features of the SDL language.

REFERENCES

[1] CCITT. *Recommendation Z.100: Specification and Description Language SDL*, volume X.1-X.5. CCITT, 1988.

[2] CCITT. *Recommendation Z.100: Specification and Description Language SDL*, volume X.R25-X.R32. CCITT, 1992.

[3] W. Glunz, T. Kruse, T. Rössel, and D. Monjau. *Integrating SDL and VHDL for System-Level Hardware Design.* In CHDL 93 - Computer Hardware Description Languages and their Applications, Ottawa, Canada, April 1993.

[4] W. Glunz and G. Venzl. *Using SDL for Hardware Design.* In Proceedings of the Fifth SDL Forum, Glasgow, 1991.

[5] T.B. Ismail, M. Abid, and A. Jerraya. *COSMOS: A Codesign Approach for Communicating Systems.* In Co-Design, Computer Aided Hardware/Software Engineering. IEEE Press, 1994.

[6] A.A. Jerraya and K. O'Brien. *SOLAR: An Intermediate Format for System-Level Modeling and Synthesis.* In Co-Design, Computer Aided Software/Hardware Engineering. IEEE Press, 1994.

[7] B. Lutter, W. Glunz, and F.J. Rammig. *Using VHDL for Simulation of SDL Specifications.* In Proceedings of the EURO-VHDL, 1992.

[8] K. O'Brien, T.B. Ismail, and A.A. Jerraya. *A Flexible Communication Modelling Paradigm for System-Level Synthesis.* In Intl. Workshop on Hardware-Software Co-Design, 1993.

[9] The Institute of Electrical and Electronic Engineers. *IEEE Standard VHDL Reference Manual.* IEEE, 1987.

[10] D.L. Perry. *VHDL.* McGraw-Hill Inc, 1991.

[11] O. Pulkkinen and K. Kronlöf. *Integration of SDL and VHDL for High-Level Digital Design.* In Proceedings of the EURO-VHDL, 1992.

6

FROM LOTOS TO VHDL

C. Delgado Kloos, A. Marín López,
T. de Miguel Moro, T. Robles Valladares

*Depto. Ingeniería de Sistemas Telemáticos, ETSI Telecomunicación,
Universidad Politécnica de Madrid, Madrid, Spain*

ABSTRACT

This paper presents the formal description technique LOTOS and analyses how it can be used as a language for describing systems abstractly. The differences between a specification language such as LOTOS and a hardware description language such as VHDL are discussed. A translation model from LOTOS to VHDL is presented.

6.1. INTRODUCTION

Seen from a historical point of view, the languages for modelling and describing hardware have been continuously increasing in abstraction level: differential equations, to boolean algebra, to register-transfer level descriptions, VHDL. VHDL allows systems to be described in an abstract manner. This trend towards increasing detail in the description of systems has the advantage of relating new concepts back to existing elements.

Nevertheless, it is instructive to take a look at other branches of computer science in which a top-down approach has been followed. By comparing these two classes of approaches, one can find to what extent the bottom-up kind of language is missing generality and elegance, and to what extent the top-down kind of language is lacking in relation to implementation aspects or, in the case of electronic design, how they can be used to design hardware.

There are many approaches to the specification of complex systems. In this paper, we will concentrate on the family of process algebraic languages, and in particular on one language in this family, namely LOTOS. LOTOS and the calculi on which it is based, namely Hoare's CSP (communicating sequential processes [9,10]) and Milner's CCS (calculus of communicating processes [16,17]), have been designed with very simple but powerful primitives with elegant mathematical properties. This is why they are called process algebras: the algebraic properties of the operators drove the design of these calculi. We can cite Milner when he offers a theory of communicating processes "for people who would like to understand them in terms of a few *primitive* ideas" or Hoare when he explains that the objective of his research into communicating processes has been "to find the *simplest* possible mathematical theory" with a number of desired properties.

Some languages, such as the process-algebraic calculi just mentioned for concurrent programming or modern functional languages for sequential programming, were designed with this simplicity in mind. Others have been given a clean mathematical semantics a posteriori. These semantics relate language constructs to mathematical concepts. It is only recently that the problem of giving a formal semantics to VHDL has been addressed [5].

In this paper, we present a process algebraic specification language, namely LOTOS, and give a short introduction to VHDL. We then compare these two languages. Finally, we sketch a translation from LOTOS to VHDL. A compiler has been implemented following the outline presented here. We will conclude with some applications of such a translation procedure.

6.2. LANGUAGES

6.2.1. LOTOS

LOTOS (Language Of Temporal Ordering Specifications) was developed within ISO (International Standardization Organization) for the formal specification of open distributed systems, and in particular for the OSI (Open Systems Interconnection) computer network architecture. The language is nonetheless applicable to distributed, concurrent systems in general, such as for instance to the description of digital hardware. Its behaviour description part was based on process algebraic ideas, predominantly on CCS, but it also borrowed some ideas from CSP.

LOTOS was developed from 1981 to 1986 and became an international standard (IS 8807) in 1989. Since then, activity around LOTOS has been steadily increasing. Tools have been built, and in particular, telecom companies are adopting it for the specification of services and protocols. But LOTOS is not only useful in the telecommunications world. Due to its generality, inherited from CCS and CSP, all kinds of concurrent systems can be defined in a formal and precise way.

In this section we will briefly describe the fundamentals of LOTOS. We will not go into all the details. The interested reader is referred to [4,13].

6.2.1.1. Actions

One of the key ideas in LOTOS (as well as CCS and CSP) is the notion of indivisible interaction. An interaction, or simply action, is the primitive concept when modelling concurrent systems. It defines the granularity for modelling. An action might be the insertion of a coin, the lighting of a lamp, the arrival of a letter. When occurrences which take time are to be modelled, one can use two actions: one for the start of the occurrence and one for its finalization.

Concurrent systems are either primitive or can be described as subsystems that interact with each other and with the environment. These subsystems could possibly also be composed of further subsystems, thus having a hierarchy. The parallel composition of subsystems is as simple as the sequential composition of statements in an imperative programming language.

Another key idea in LOTOS is the abstraction from causality. When two concurrent systems interact, it is immaterial when modelling with LOTOS which of them produced the action and which of them is receiving it. There is a symmetry: both systems agree in producing an action, it just happens due to the capabilities of both systems. There is a similarity with systems in nature – the moon moves around the earth, due to some properties both systems have (gravitational forces) – or in electricity – a circuit composed of say resistors, capacitors and coils behaves in one particular way due to the characteristics of its components, the way they are interconnected and the initial conditions. The same occurs in LOTOS when composing concurrent systems: each system has some capabilities; the interconnection and these capabilities determine the behaviour of the whole. But note that the interaction in LOTOS is discrete.

Thus, an action can be identified with the interaction point (called gate in LOTOS) on which the action takes place. We only need to name gates and can use the same name for the action that takes place on it. Each system has a fixed interface consisting of a number of gates with given names. In Figure 1 we can see the difference between the interface of the system and the interfaces of its processes. It is possible to rename the gates.

When two systems agree to interact, i.e. synchronize on a gate, they might exchange some values. Communication is only possible through interaction at common gates. A gate may offer a value to another gate that synchronizes with it. This models output. The syntax is `g!5`, for gate g outputting value `5`. A gate may accept a value from another gate, to be stored in a variable. This models input. The syntax is `g?a:integer` for a gate `g` accepting an integer to be stored in variable `a`. Figure 2 illustrates this interaction.

But the LOTOS communication mechanism offers more possibilities than the transfer of a value from one system to another one. It is also possible that more than two systems synchronize on a gate. This is called *multi-way rendezvous*. Accepting a value (or offering a variable, i.e. input) can also be understood as offering all the values of the corresponding type. All the systems synchronizing on a gate have to agree on a value compatible with all the values offered.

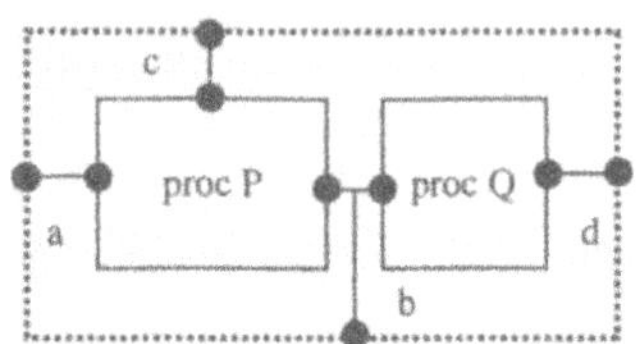

Figure 1: System Interface

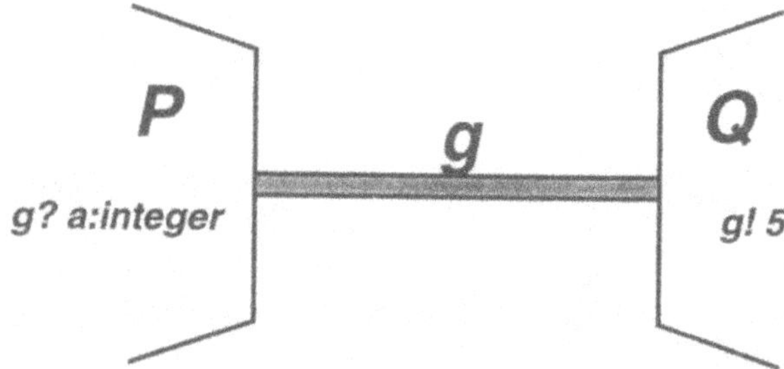

Figure 2: Communication through a Gate

Examples: Let `E` and `E'` be expressions of type `type`

- `g!E` and `g?a:type` can synchronize leading to *value passing.*
- `g!E` and `g!E'` can synchronize provided that the values of `E` and `E'` coincide. This is *value matching.*
- `g?a:type` and `g?b:type` can synchronize leading to a *value negotiation and generation.*

6.2.1.2. Behaviour Expressions

The behaviour of concurrent systems in LOTOS is described by means of behaviour expressions, in the same way as arithmetic expressions describe numeric computations. In this section we will describe the basic constructs to describe behaviour expressions.

First, there is the system that cannot perform any action. It is called `stop`. This is the inaction or deadlock system. It serves to define other systems that eventually stop. There is also another system that does nothing except finalizing successfully. This is called `exit`. It is related to `stop`, since it can also be seen as a system that offers an action of success and then stops.

There are two dimensions to consider: sequentialization and choice. With the *prefix* operation it is possible to define a system that first is capable of performing an action and then behaves like another system. The symbol for this operation is ;. For example, let a and b be actions, then `a; b; stop` is the behaviour of a system that first is able to perform action `a`, then `b`, and then stops. `B1 [] B2` is a *choice behaviour expression* which behaves either like B1 or like B2. Once an action is chosen from one component, the other component disappears from the resulting expression. Both operations can be composed as in the following example:

```
a; (c; stop [] b; stop)
[]
b; a; (a; exit [] c; exit)
```

Behaviours can be represented graphically by the so-called *synchronization trees*. The behaviour expression from the example above would be represented as in Figure 3.

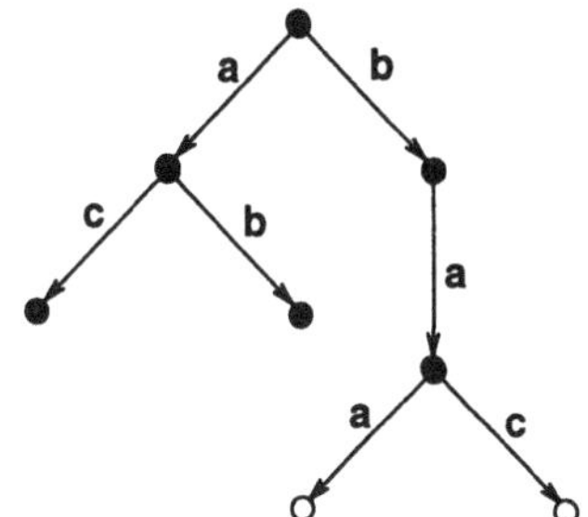

Figure 3: Synchronization Tree

6.2.1.3. Processes

In the same way that an arithmetic expression can be abstracted to a function, a behaviour expression can be abstracted to a process. A process is a behaviour expression parameterized on the names of gates. It can also depend on additional arguments. The abstraction of a behaviour expression to a process also allows the definition of (directly or mutually) recursive behaviours.

```
process P [in1, out1] (n: integer) noexit :=
in1?v:integer; P(v)
[]
out1!n; P(n)
endproc
```

This is a process that either accepts an integer at gate `in1` and reinstantiates itself with the integer accepted, or offers the argument `n` at gate `out1` and reinstantiates itself with the same argument. This process represents a one place memory or buffer.

6.2.1.4. Concurrency

It should be noted that all behaviour expressions describable in LOTOS can be defined using the four basic operators (`stop`, `exit`, `;`, and `[]`) together with recursion. This style of describing systems is called monolithic. It is nevertheless often useful to describe systems as a composition of subsystems in parallel, not just because this might reflect the natural decomposition of systems. It is simpler to say that a system is composed of several systems interacting concurrently, rather than describing all the possible ways in which these subsystems might interact.

To this end, LOTOS has several operations for expressing parallel composition:

- Interleaving `|||`: Concurrent composition without synchronization
- Total synchronization `||`: Concurrent composition with synchronization on all common gates
- Selective synchronization `|[...]|`: Concurrent composition with synchronization on the given gates (general case)

The *interleaving* of two behaviour expressions allows all possible combined traces of actions respecting the traces of the individual components. Example: `in1; out1; stop ||| in2; out2; stop` is equivalent to

```
in1; (out1; in2; out2; stop
   [] in2; (out1; out2; stop [] out2; out1; stop))
[]
in2; (out2; in1; out1; stop
   [] in1; (out1; out2; stop [] out2; out1; stop))
```

This equivalence can be checked more clearly in Figure 4.

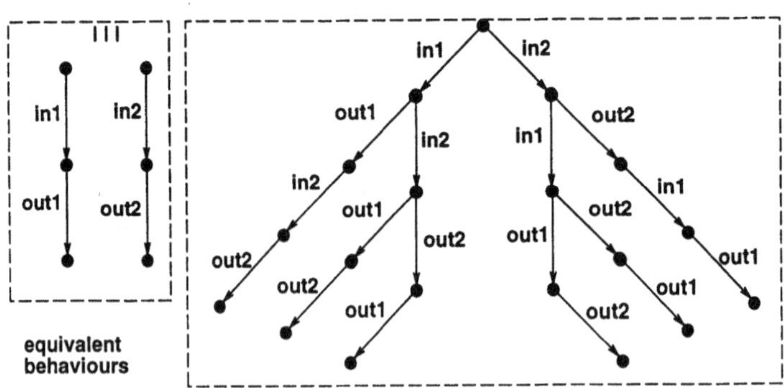

Figure 4: Equivalent Trees

In *total synchronization* both behaviour expressions synchronize on all their common gates. One cannot continue without the other when common actions are encountered. On independent actions no constraint is imposed.

Example: `in1; out1; exit || (in1; out2; exit [] in1; out1; exit)` is equivalent to `in1; stop [] in1; out1; exit`. Depending on a non-deterministic choice, the behaviour might lead to deadlock or finish successfully.

Selective synchronization is the most general case, since it includes the two previous ones. The list of gates on which the two behaviours shall synchronize is explicitly given. If this is the empty list, interleaving is obtained, and if it is all common gates, total synchronization is obtained.

6.2.1.5. Hiding

As we said before, hierarchy is a very important concept in LOTOS. It is important to structure complex behaviours. The process abstraction concept allows parameterized behaviour expressions to be encapsulated. But one normally does not want to see the gates of all the interacting processes at the interface level. The *hide* operator allows actions to be made invisible. The actions are internalized so that no behaviour expression can interact with them. Nevertheless, the internalized action can have an effect on its future behaviour. When an action is hidden as in `hide a in a; b; stop [] a; c; stop` the resulting behaviour expression has in its (`a`'s) place the so-called internal action `i`. In the example above, the behaviour expression would be equivalent to `i; b; stop [] i; c; stop`. The internal action is special and cannot synchronize with any action.

6.2.1.6. Enabling and Disabling

In LOTOS, the passing of control among processes is described by two operators: *enable* and *disable*. Sequential composition is modelled by the *enable* operator. `B1 >> B2` means that behaviour `B1` is followed by behaviour `B2` if and only if `B1` has terminated successfully (*exit*).

The *disable* operator models interruptions and exceptions to the normal behaviour of systems. For example:

```
Normal_behaviour [> Exception_handling
```

If any of the events of `Exception_handling` occur, the execution of `Normal_behaviour` will be interrupted. This interruption can occur until successful termination of `Normal_behaviour`.

6.2.1.7. Data

Little has been said about the data types. We have just assumed that there were some data types available for the values transmitted through gates. The mechanism to define and to deal with data types is based on ACT ONE [7]. The definition of a data type is composed of two parts: *signature* and *equations*.

The signature defines the set of values and the operations to work with these values. That makes a difference with programming languages where types and operations (functions and procedures) are defined separately.

Equations give the semantics to the operations defined in the signature. They define the properties that the operations have to suffice. No implementation is needed for the operations. In fact all the implementations are valid, as long as they satisfy the properties. That makes the second difference with programming languages where the semantics of the operations is predefined in the language or is given by the implementation of the operation itself.

LOTOS has not predefined types, but it allows the use of predefined libraries. These libraries contain the meaningful types for application domains. Nevertheless, new types can be defined in a specification.

The following is an example of definition of the boolean data type:

```
TYPE Boolean IS
  SORTS         (* beginning of signature *)
    Bool
  OPNS
    true, false: -> Bool
    not: Bool -> Bool
    _ and _, _ or _: Bool, Bool -> Bool
                (* end of signature *)
  EQNS
    FORALL x, y: Bool
    OFSORT Bool
      not (true) = false ;
      not (false) = true ;
      x and true = x ;
      x and false = false ;
      x or true = true ;
      x or false = x ;
  ENDTYPE
```

6.2.2. VHDL

VHDL originated from the VHSIC (Very High-Speed IC) Program, funded by the US Government DoD. Developed by IBM, Texas Instruments, and Intermetrics, all the rights to the new language were transferred to the IEEE in March 1986. In December 1987 it was adopted as the first industry standard hardware description language. A reaffirmation of the standard took place in May 1993, leading to the publication of the IEEE Std. 1076-1993 [11]. The reasoning behind this re-standardization process is the intention of the IEEE to maintain VHDL in good health and in suitable condition for industry's needs, thus encouraging its use by the design community.

The VHDL language was intended to couple with designs at different levels of abstraction. Its use ranges from the description of the structure of a system to the modelling of a discrete gate. This fact makes it possible to use VHDL under a large set of different tools through the design cycle. The concept of hierarchy is behind this capability of describing hardware behaviour at such different levels.

Another important issue of VHDL is the possibility of simulating descriptions. This is the core of the language. Up to now, the simulation semantics has been the only standard semantics for VHDL, while the synthesis semantics still depends on the synthesis tools. The idea of WYSIWYS (What You Simulate Is What You Synthesize) is present, but a standard synthesis semantics is needed.

The language itself is widely treated by several books [2,8,3]. This section only remarks on some aspects of the language relevant to the purposes of this paper.

Concurrency in VHDL is achieved via the parallel execution of sequential statements (processes). The execution is governed by the events queue, which stores and updates *projected* waveforms for the different signals. Processes suspend and resume execution at certain points where explicit or implicit *waits* are placed. Suspended processes resume execution when an expected event takes place. A process may be resumed by several different events. For a detailed description of the mechanisms involved (simulation cycle, sensibility lists, signal assignments, waits, etc.), see the suggested bibliography.

The hierarchy in VHDL comes from the notions of entity and component. They basically describe the interface of a device, in order to be bound in a concrete implementation of a design. An architecture describes the behaviour (or the structure) of an entity. Different architectures can be defined for the same entity and bound to different components instantiated from a higher-level hierarchy architecture. This is the main point that enables the use of VHDL in different levels of the life-cycle of a design.

The last point we want to mention for our purposes is the package STD_LOGIC_1164, developed by the IEEE Model Standards Group (PAR 1164).This package defines a standard to be used by designers for describing the interconnection data types used in VHDL modelling.

6.2.3. LOTOS vs VHDL

6.2.3.1. Specifying vs. Programming

What is the difference between specifying on the one hand and programming with a programming language or modelling with a hardware description language on the other hand? In general terms, a specification can be more abstract, it can describe properties in a non-constructive way, but is not restricted to that. For instance, one might specify a matrix inversion function as follows:

```
inv: matrix  → matrix
inv(a) × a = I
a × inv(a) = I
```

Analogously, a list inversion function might be specified as follows:

```
inv: list → list
inv(emptylist) = emptylist
inv(elem & l) = inv(l) & elem
```

It is clear that the matrix inversion specification is non-algorithmic, whereas one can immediately derive a program from the list inversion specification. Thus, a specification can be so abstract that it is hard to derive an algorithm from it directly, or it can be as concrete as a program, or it can be executed, but perhaps with low efficiency.

Another view that is not inconsistent with the previous one arises when considering the design of systems as a hierarchy. In this view, implementations are concretions of specifications. In a layered design model, the implementations of one level are the specifications of the level below.

In Sections 6.2.3.2. and 6.2.3.3. we compare both languages. The comparison is made bearing in mind the purpose of establishing a translation model. We will compare at two levels: structure and behaviour. The reason for separating the analysis is because VHDL has different mechanisms for description at register-transfer level and algorithmic level. But LOTOS does not differentiate between the specification of structure and behaviour.

6.2.3.2. Description of Structure

Both VHDL and LOTOS can express hierarchy. Hierarchy plays a key role in the description of systems. In fact, both languages use hierarchy as a mechanism for traversing the abstraction/refinement ladder. Modularity, reusability, documentation and maintenance are also achieved through the proper use of hierarchy.

Let us see what abstraction and refinement mean in LOTOS. It is a formal language for specifying systems. Systems can be described in LOTOS at very different levels of abstraction. The refinement process will introduce implementation details, eliminating high-level structures in the specification. As the abstraction level of the specification decreases, the implementation task becomes easier. Different implementations of the same specification are possible. The purpose of these implementations is to help the user to understand the system. This is achievable with the LOTOS specification environment, which offers the possibility of executing the model. This leads to early prototypes of the specification. The evaluation is very helpful for making design decisions and for completing the requirements capture task. Specifications are refined with the help of these implementations.

Systems can also be described at different levels of abstraction in VHDL. The existence of a hierarchy allows the integration of modules that are described at different levels of abstraction. In VHDL, every description is simulatable, and implementation details are also added in the refinement process. The level of abstraction and refinement achievable in VHDL is lower than in LOTOS, where several interpretations of the specifications can be given, and the implementation is done in a different language.

In addition to this, there exist more features to handle design units, packages and libraries in VHDL. Blocks, components, configuration and use clauses are powerful elements of the language for building design environments in which several teams of designers can work. This is a difference with LOTOS. Only data type libraries can be described in LOTOS. LOTOS has no name handling mechanisms like VHDL, which allow one to define complex design environments.

Another point at which LOTOS and VHDL differ is in the interconnection of the building units. In both languages, channels are established by interconnecting the interfaces of the components. However, such interconnections are all identical in VHDL, while they can be given different semantics in LOTOS. Interleaving, total synchronization and selective synchronization are the operators that LOTOS allows for the interconnection of the components.

One more difference concerning the interface is the type of the channels. VHDL imposes a rigid schema in which the channels are typed, and only ports of the same type can be interconnected. LOTOS does not impose any limitation on the gates. At the same gate and in different synchronizations, different event structures may be used. Moreover, LOTOS has a mechanism of relabelling the gates of every behaviour unit. This is not possible in VHDL unless the ports have the same type.

The last point of difference with respect to structure lies with the parameterized structures. The possibility of describing and using these exists in both languages, but the expressiveness of LOTOS is greater. VHDL allows the use of generics and the configuration of parameterized structures by means of the generate statement. In LOTOS such structures are possible, but dynamically configurable structures can also be described, i.e. the structure may be unknown at compilation time and only known at run time.

Concepts	**LOTOS**	**VHDL**	**Compatibility**
Data	ADT	high level	restricted
System Components	processes	entities	ok
Comp. Interface	gates	ports	ok
Comp. Interconnection	parallel op.	wiring	restricted
Comp. Instantiation	dynamic	fixed	restricted

Table 1: Compatibility of Description Capabilities

Table 1 covers the compatibility of LOTOS and VHDL for describing the architecture of the system.

6.2.3.3. Description of Behaviour

The capability for describing behaviour can be analyzed at two levels: *communication* and *data*.

6.2.3.3.1. Communication

There are three main points to deal with when analyzing communication: *events*, *processes* and *synchronization*.

event

The concept of event exists in both languages, though with different meanings.The LOTOS event is the atomic unit of interaction. Only one event is allowed to occur at one instant, though they can be as close or as far apart as desired, in an interleaved way. Another point is that events can be used in LOTOS for interchanging several sets of data values, even simultaneously. Original LOTOS only allows the definition of event sequences with a timed ordering of events. We have enhanced LOTOS semantics with time properties, and the T-LOTOS language (see [19]) allows the definition of time intervals associated with events.

An event in VHDL is defined as a change of value of a signal. Since the definition of VHDL is based on a discrete simulation algorithm, events trigger the computation of new simulation cycles. Two events can occur at the same physical time but at different simulation-cycles. This is called a *delta-delay*. Several events can occur in the same instant. In this sense, VHDL events can occur simultaneously, although the events in VHDL are simple, i.e. a driver can only be given a simple value.

process

Both languages use the notion of process, but also with different meanings.The behaviour part in LOTOS is defined by a set of processes combined via binary operators. A run of a LOTOS process may be seen as a sequence of actions, some of which require synchronization with the exterior. A process evolves independently of other processes when no synchronization is required. The gates involved in the synchronization are defined in the header of the process. Parallelism may exist inside a LOTOS process, as parallel operators can appear inside its body. LOTOS operators define two classes of relation among processes: passing of control and synchronization. The passing of control performed by the *enabling* operator is semantically equivalent to resuming a suspended VHDL process. The disabling operator describes an interruption without control return. The *disabling* operator also has a direct implementation in hardware. The interpretation of the passing of control is described in the next section. LOTOS does not impose a limit on the number of processes currently running.

A VHDL process describes a sequential behaviour. Processes execute concurrently. A process can require synchronization with other processes, but the synchronization is much more restricted than in the case of LOTOS. Synchronization is done only by checking changes on signals and predicates over their values. The signals involved in the external synchronization are defined in the header of the process. Parallelism cannot exist inside a process as it does in LOTOS. There is no limit to the number of processes currently running, but it will be known and finite at compilation time.

synchronization

The LOTOS synchronization mechanism is based on multiway rendezvous. This implies the synchronization of *n* processes, and the simultaneous occurrence of the *n* events (one for each process). The processes participating in a synchronization can negotiate or even invent the value or set of values that will constitute the event. All this work has to be done atomically, i.e. either all the processes perform the same action (as the result of the commitment of the *n* events), or none of them completes the event. The execution of one LOTOS event implies: synchronization point + value interchange + *n* partners. LOTOS allows the evaluation of expressions through *let* statements.

This model is quite different from the wire interconnection with value sending of VHDL. The translation is possible in practical cases, but it is not straightforward. It will require, in the general case, the definition of a complex signal interchange protocol.

Concepts	LOTOS	VHDL
Value Evaluation	ok	ok
Value Interchange	n values	1 value
Value Transmission	value negotiation	explicit

Table 2: Synchronization Possibilities

Table 2 shows that both languages provide value evaluation. The differences are related to values. While LOTOS allows the negotiation of *n* values among *m* processes for each event, VHDL restricts the interchange to one value for each signal. For our translation purposes the key elements in the previous table will be the synchronous interpretation of LOTOS and the *n* value negotiation among *m* partners required by the execution of one event.

When VHDL describes the sending or reception of a value, it does not impose any condition on the corresponding process. It has a direct hardware interpretation: if the process is sending a value, it puts a value on a wire, then it forgets the value. It is the responsibility of the counterpart process to get this value before it is changed. On the other hand, the process that gets a value takes the value from a wire, without worrying about the process that put the value.

This synchronization is more restrictive than the LOTOS rendezvous. They are not incompatible, and the LOTOS synchronous interpretation may be implemented using a protocol as discussed in the next section.

Table 3 summarizes the semantic properties of the three synchronization models: 1-*n* for LOTOS 1 to *n* value passing and VHDL for VHDL value sending. We assume $n \geq 0$.

Table 3 shows how the 1-*n* value-passing model suits the VHDL semantics much better. The only relevant difference between these models is the number of receivers of the value. This problem will be solved by the translation model without imposing any extra limitation. The exact definition of this translation model is realized in the next section.

Concept	**Rendezvous**	**1-n**	**VHDL**
Partners	n	n+1	n+1
Value Provider	≥ 0	1	1
Value Receiver	≥ 0	n	n
Value Generation	yes	no	no

Table 3: Synchronization Models

6.2.3.3.2. Data

The data types used in both languages are very different. The data part of LOTOS is based on ACT-ONE [7] for the algebraic specification of *abstract data types*. ADTs do not imply any particular data implementation. Furthermore, in the general case they are not implementable as common high-level programming language data types. An ADT's semantics is defined by means of equations.

VHDL data types are similar to data types of common imperative high-level languages. Nevertheless, VHDL synthesis tools generally significantly restrict the allowed set of data types.

ADTs are also the concept that makes it possible to exchange, even simultaneously, several sets of data values through the same synchronization point. ADTs define the semantics of the values, but do not impose an implementation. That implies the possibility of defining both realizable and unrealizable data values.

Due to this difference in the definition of data types, it seems unsuitable to define a transformation tool for automatic translation of LOTOS data types to standard VHDL.

The problem will be solved by identifying the translatable data types and restricting the data types and data values to that set of translatable elements. The next section defines which ADTs may be used in LOTOS, and how to translate them into VHDL data types.

6.3. TRANSLATION

In this section, we discuss how we translate from LOTOS to VHDL. Section 6.3.1. deals with aspects of the translation which determine the target architecture. In Section 6.3.2. we deal with synchronization aspects. The translation of the data types is discussed in Section 6.3.3. Finally, in Section 6.3.4. the translation subset is presented.

6.3.1. Architectural Aspects

The order of the discussion will be the following: processes, gates, parallel operators, architectural considerations and other operators.

6.3.1.1. Processes

The expression of hierarchy in a design is natural in both languages. LOTOS processes are the basic units used to build up hierarchy. They execute concurrently, synchronizing as stated when they are instantiated. They can express sequential behaviour as well as concurrently instantiate in any synchronization scheme needed by the designer.

There are four elements of the VHDL language that can be used to express concurrency: processes, subprograms, blocks and entities. They can all be a target for LOTOS process mapping. However, they are quite different when considered from the point of view of expressing hierarchy.

- VHDL processes can activate others by means of events on signals. However, in VHDL, processes cannot include processes, and so they do not express the notion of hierarchy. They cannot be a possible target mapping of a general LOTOS process[1].
- Subprograms, when called from within a block or an architecture, execute concurrently, and they can call others, but always sequentially. This prevents us from using them as targets for the mapping of a general LOTOS process. Only the first call executes in parallel.
- Entities apparently describe only the interface to a device, but they are the key to expressing hierarchy. They can instantiate other entities in any wiring needed (providing it is correct). This is done using the notion of a component.
- An entity defines an external block, and both blocks and entities give us the same capability of expressing hierarchy and concurrency.

Following this analysis, we conclude that LOTOS processes can be mapped onto blocks or entities. We will translate them into entities.

[1] Readers familiar with Verilog know that inside a sequential behaviour a parallel one can be defined by means of the *fork* statement. No statement like this exists in VHDL.

6.3.1.2. LOTOS Gates

Communications take place at LOTOS gates. The word communication should be interpreted at a high level (i.e. LOTOS gates offer sets of values of different types, upon which the synchronizations may occur).

Signals connect components, blocks, processes, etc. They transport the values according to their type and mode. The role of the *events queue* and the delays (transport and inertial) provided by VHDL, though defined in the simulation semantics, are not always clear in a synthesis environment, where designers are not allowed to use time (at high level).

Signals can be used as value drivers, but they are not defined as synchronization points. When both aspects are required, they are not enough to translate LOTOS gates. In the general case a LOTOS gate should be translated by three signals and a library component performing the synchronization. We show this in the discussion of synchronization. However, LOTOS gates could be implemented directly with signals for certain purposes and under some conditions.

Let us illustrate the previous statements with an example. The example is a processor which cooperates with other devices to perform some work. The cooperation can be modelled as communication and synchronization on different gates. The processor has to exchange information about data, memory addresses, and operations to perform on the different devices. It also has to allow non-maskable and maskable interruptions. At this level of abstraction, the header of the process in LOTOS would look like this:

```
PROCESS processor [dir,data,op,irq,nmi] := NOEXIT
(* behaviour *)
ENDPROC
```

And in our translation we would obtain:

```
ENTITY processor IS
  PORT(
    Transl_decl_gate(dir);
    Transl_decl_gate(data);
    Transl_decl_gate(op);
    Transl_decl_gate(irq);
    Transl_decl_gate(nmi));
END processor;
ARCHITECTURE beh OF processor IS
  Transl_beh_process(processor);
END beh;
```

The ports of the entity have not yet been translated. The reason for this is that no decision about the interpretation of the LOTOS gates has been made. In the general case, in our model, they would be translated by three signals (one for the data and two for the synchronization protocol). The mode and type of the signals are still unknown, as in the specification it is only said that some LOTOS gates exist in the LOTOS process, but nothing is said about the synchronizations that are to be performed at the

LOTOS gates. With information about the interactions that occur in gate `data`, `Transl_decl_gate(data)` will be translated by:

```
data_a: mode_a std_ulogic_vector(1 downto 0);
data:   mode std_ulogic_vector(size_ut downto 0);
data_r: mode_r std_ulogic_vector(1 downto 0);
```

The modes of `data_a` and `data_r` depend on whether `data` is used to output data or to input data. `size_ut` stands for the size of the bitvector equivalent to type *ut* (undefined type). Similarly, `size_myinteger` would stand for the size of the bitvector equivalent to type `myinteger`. The ports `XXX_a` and `XXX_r` arise due to the asynchronous nature of LOTOS; this is discussed in detail in Section 6.3.2 . During specification, we can get to a different port list without ports `XXX_a` and `XXX_r` for some of the LOTOS gates. This could be achieved whenever the synchronizations that take place in these gates are synchronous, i.e. they are specified in terms of a clock[1]. The discussion of synchronization and the introduction of a clock at the LOTOS level of abstraction can be found in Section 6.2.3 .

6.3.1.3. Parallel Operators

There are three binary LOTOS operators to compose processes in parallel: interleaving, full synchronization and selective synchronization. The synchronization scheme depends on the gate lists of the processes involved. Let *P* and *Q* be two LOTOS processes with gate lists *g1* and *g2*, respectively. Let $I=g1 \cap g2$ be the intersection of the lists and $D=(g1 \cup g2) \backslash (g1 \cap g2)$ be the union of the gates of *g1* and *g2* that do not belong to *I*. *P* || *Q* (full synchronization) will deadlock if an offer is made at some gate in D. *P* ||| *Q* (interleaving) will ***purely*** interleave events at gates in *D*, but ***non-deterministically*** interleave events at gates in *I* (it is not predictable whether *P* or *Q* will succeed in an offer that both could accept). Let $g3 \subset g1 \cup g2$, *P* | [*g3*] | *Q* (selective synchronization) is equivalent to synchronize *P* and *Q* in gates in *g3* and interleave them in the rest of the gates. To summarize, we are dealing with three possible interactions: pure interleaving (no interaction), non-deterministic interleaving, and synchronization.

If the interaction is not pure interleaving, these operators need special components to be translated. The case of synchronization requires the agreement of the two involved behaviours on every phase of the synchronization. This can be performed with simple AND and OR components. Non-deterministic interleaving is much more difficult, as VHDL is deterministic. Even leaving non-determinism apart, the problem is complex, and there exists no distributed algorithm that effectively solves the general case. However, some restricted forms of interleaved synchronization can be solved, and they only require a component with some memory to ensure fairness. The restrictions refer to the direction of the flow of information, that should be fixed and of a certain type, to avoid large (possibly infinite) computations. One centralized implementation of the general case is shown in [14]. The component initially checks both behaviours in turn for synchronization. The protocol begins when one of them wants to synchronize.

[1] Data dependencies could also exist.

When the synchronization ends (successfully or not), the component memorizes which of the two behaviours was involved (to avoid deadlocks). Non-determinism is lost at component instantiation, and one of the behaviours is favoured because it is checked before the other.

To continue with the previous example, let us specify a memory component which communicates with the processor. The memory and the processor agree on a direction, on the operation to perform, and on the data. This synchronization is modelled in LOTOS as:

```
processor[dir,data,op,irq,nmi]
|[dir,data,op]|
memory[dir,data,op,refresh]
WHERE
  PROCESS memory [dir,data,op,refresh] (data_arr:integer): NOEXIT:=
   (* behaviour *)
  ENDPROC
```

Note that the two LOTOS processes synchronize in gates `dir`, `data`, and `op`, and interleave their behaviours in gates `refresh`, `irq`, and `nmi`. According to the discussion on LOTOS parallel operators, this behaviour is mapped onto separate components, one for each gate, in which LOTOS processes synchronize as follows:

```
processor1 : processor PORT MAP(
        dir_a => dir1_a, dir => dir1, dir_r => dir1_r,
        data_a => data1_a, data => data1, data_r => data1_r,
        op_a => op1_a, op => op1, op_r => op1_r,
        irq_a => irq1_a, irq => irq1, irq_r => irq1_r,
        nmi_a => nmi1_a, nmi => nmi1, nmi_r => nmi1_r);
memory2  : memory PORT MAP(
        dir_a => dir2_a, dir => dir2, dir_r => dir2_r,
        data_a => data2_a, data => data2, data_r => data2_r,
        op_a => op2_a, op => op2, op_r => op2_r,
        refresh_a => refresh2_a, refresh => refresh2,
        refresh_r => refresh2_r);
Synch3: synch_op PORT MAP(
        inp_a => dir_a, inp => dir, inp_r => dir_r,
        out1_a => dir2_a, out1 => dir2, out1_r => dir2_r,
        out2_a => dir1_a, out2 => dir1, out2_r => dir1_r);
Synch4: synch_op PORT MAP(
        inp_a => data_a, inp => data, inp_r => data_r,
        out1_a => data2_a, out1 => data2, out1_r => data2_r,
        out2_a => data1_a, out2 => data1, out2_r => data1_r);
Synch5: synch_op PORT MAP(
        inp_a => op1_a, inp => op1, inp_r => op1_r,
        out1_a => op2_a, out1 => op2, out1_r => op2_r,
        out2_a => op_a, out2 => op, out2_r => op_r);
```

The components `Synch_X: synch_op` are library modules and have been carefully tested. They perform the synchronization operation of the ports `out1` and `out2` with port `inp`. Analogously, there exists a `component interl_op` which performs the interleaving synchronization. Note that in the previous example it is not required, because the processor and the memory synchronize at every common gate, that is a way to avoid the existence of non-determinism.

6.3.1.4. Architectural Considerations

Up to this point we have discussed the aspects of the translation that deal with concurrency. LOTOS processes and parallel operators, in their most general form, need to be mapped onto entities. The rest of the operators we discuss in this section can be translated within the body of VHDL processes defined by the mapping of LOTOS processes and parallel operators. The target VHDL processes can be described in very different ways. However, finite state machines (FSM) are the most common architecture to be treated by synthesis tools. We use FSMs to describe sequential behaviour. The remaining LOTOS operators (apart from synchronization) can be expressed in one or more states of the FSM. The code obtained in this way is suitable as the input to a commercial synthesis tool. Some work is still left for the algorithm level synthesis, because the data (operations and store) are inside the FSM. From this point of view, the FSMs obtained in this manner are more than pure controllers. They deal with the communications part of the design. We discuss data manipulations in Section 6.3.3.

To sum up, the parallelism of the VHDL obtained resides in several processes that can be active at the same time.

The FSMs we use have an asynchronous common reset for initiation purposes. They also have an *enable* input to make the modelling of *instantiation*, *enable* and *disable* easier. They are composed of two processes: a sequential and a combinational one. Several descriptions exist of this modelling of FSMs in VHDL. Readers are referred to [1,18].

6.3.1.5. Other Operators

Sequential behaviour can be easily described in a FSM, as every state consumes at least one clock-cycle. Let us explain the mapping for the following LOTOS operators:

action prefix
It expresses sequentiality, and so implies that one behaviour should occur before the other. We express this in VHDL by partitioning the two behaviours into multiple states, the last state of the first behaviour enabling the first state of the second one.

choice
A non-deterministic LOTOS choice must be mapped into a deterministic selection in VHDL, because VHDL is deterministic. There are several possibilities for expressing alternative sequential behaviour in VHDL. We use the *if else* statement.

stop
Several means to express LOTOS inaction, i.e. deadlock, exist in VHDL. In our target architecture we can inhibit the execution of the FSM, wait for a condition that never holds or for an event that never occurs. Our mapping consists in leaving the state of the FSM unchanged, thus preventing the process from progressing.

exit, enable
The LOTOS *exit* termination expresses sequential behaviour. The *exit* behaviour can be understood as the occurrence of an event of successful termination δ followed by a *stop*. A behaviour B2 is enabled by behaviour B1 if and only if B1 performs the δ action, i.e. it executes successfully. It should be noted that it allows us to specify that a behaviour begins (LOTOS *enabled*) when others have finished (LOTOS *exited*). In our target architecture there exists one *enable* signal per entity. That signal controls the progress of the FSM. Thus, we can map B1 enables B2 onto B1 activating B2's enable signal.

disable
Interruption is a natural concept in hardware, and it is easily modelled in LOTOS with the *disable* operator. Our FSMs have an asynchronous *reset*, which stops the progress of the FSM. B1 is disabled by B2, if the initial state of the FSM F2 resets the FSM F1, F1 and F2 being the translations of behaviours B1 and B2, respectively.

We refer the reader to the appendix for a complete example.

6.3.2. Synchronization Aspects

In Section 6.2.1. the synchronization of LOTOS was broadly explained, and the restrictions we imposed were discussed in Section 6.2.3. Let us briefly summarize the results of the discussion that deal with synchronization: synchronization flow is fixed, i.e. one of the partners always passes one value or more, while others agree with the value. LOTOS synchronization is performed by a multi-way rendezvous. VHDL performs synchronization by the use of signals. Therefore, we have to decompose the sophisticated LOTOS rendezvous in terms of the much simpler VHDL signal interchanging and pick a timing granularity suitable for the translation of simultaneous actions, value passing and value matching during the synchronization.

In our translation, we use the modularization capability of VHDL and have defined and carefully tested two components which perform the synchronization. Thus we obtain several benefits such as: the code is easier to read, these modules are synthesized and optimized independently, they offer a great capability of parallelism. The code for an offer in a given gate is equivalent to loading the correspondent module with the appropriate values, triggering the module, and sending an ok or nok. This scheme allows us to trigger several modules in parallel, which is needed in the translation of the LOTOS choice. Only one of the modules is enabled in order to commit the synchronization, while the others are reset.

The protocol used to translate the LOTOS synchronization is decomposed into three stages:

- Wait for all partners *ready to begin*.
- Data exchange (assignments, and operations to check whether it is valid for every participant).
- Commitment or abortion of the synchronization.

The previously mentioned synchronization modules, as well as the parallel composition modules behave according to this protocol.

An important aspect of the synchronization is still open: our system is designed for LOTOS, but its environment does not necessarily behave in the way that LOTOS expects. When we come to the integration, a question arises: what should the interface look like, what are the constraints? To answer this question properly, two points must be taken into consideration: first, recall what was said in Section 6.2.3. about the model of the environment in LOTOS; second, the environment may (or may not) be synchronous and knows nothing about our protocol. If the environment happens to be asynchronous, the solution is clear: our protocol should be adapted to the environment. Protocols always behave in the order request – action – acknowledge (positive or negative), this only implies an adequate wiring of the system output with the environment.

If the environment is synchronous, some special terminal components should be attached to the signals of the system. Recall that we have translated a LOTOS gate by three signals, and that there exists an asynchronous protocol behind the behaviour of these signals. Terminal components simulate the protocol (requests and acknowledges) and pass from our three-signals convention to a one-signal convention, obtaining thus an interface congruent with the environment. Of course some timing constraints should be met in order to ensure the correctness of the synchronization. In normal hardware devices, these constraints depend only on two parameters: the number of participants in each synchronization and the delay of the modules that perform the protocol, both of which are known at compilation time. In the general case, this timing constraints can depend on the data and on the state of the system, but this is due to the expressiveness of LOTOS. For our purposes we do not deal with these situations.

The introduction of this protocol looks reasonable when we are modelling asynchronous protocols. A synchronous version, where every action is performed in a particular fraction of clock cycles, makes more sense in synchronous designs. However, this depends completely on the degree of abstraction of the specification. The designer may not want to fix the model to use, but to explore the design space and evaluate the different solutions. A synchronous implementation of the specification can be obtained from an expansion of the behaviour. Let S be a specification (in LOTOS), E the environment (also in LOTOS). If we are able to transform S to get a unique (recursive) process, described in terms of action prefixes, guards and choices, and if the implementation of the environment conforms to E, we can obtain a synchronous implementation of the transformed version of S. Moreover, a FSM where LOTOS actions are translated as signal (or variables) assignments and the clock cycle is longer than the longest data manipulation would be a valid implementation.

6.3.3. Data Part

There is no predefined data type in LOTOS, and every datum is defined axiomatically by means of equations. There are some libraries defining the most often used data types, but the definition of new types usually needs long and verbose equational descriptions. In order not to reduce the capability of LOTOS to a set of supported data types in a library, but to allow the definition of new types, we have decided to map every data type to an equivalent bit vector representation. This view is reasonably close to an implementation. The bit vector domain we have chosen is the standard package developed by the IEEE, logically named STD_LOGIC_1164.

The user can define new data types in LOTOS, but they have to be defined in VHDL as follows:

- Number of bits of the bit vector representation
- Type converter functions (from/to bit vector)
- Operations defined on the type in both domains.

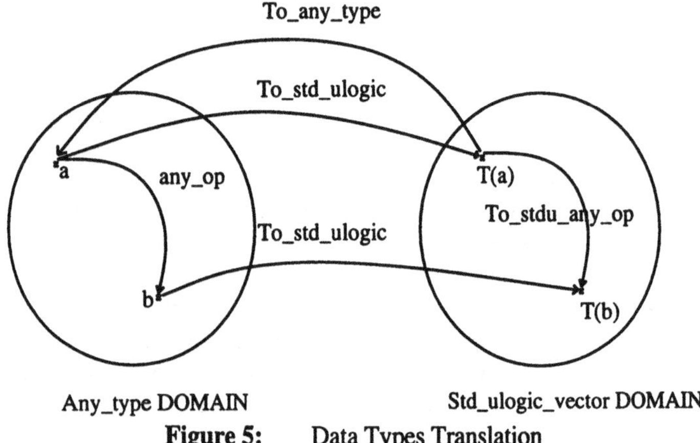

Figure 5: Data Types Translation

Figure 5 illustrates the functions needed to deal with the arbitrary type `any_name` with function `any_op1`, in the bit vector domain. To integrate this handling of data types in an automatic compiler from LOTOS to VHDL, all the functions should be invertible.

This mappings help us to deal with the problem of resolved signals that could appear when translating LOTOS processes with mutual instantiation. For instance, when process P instantiates process Q, and vice versa, and they both use a gate which offers a value (with *out* directionality), a resolution function is needed in the translation of the system. In this approach, we map the data types onto *std_logic_vectors* and use the resolution functions provided in package STD_LOGIC_1164.

6.3.4. Translation Subset

After having compared both languages and discussed how LOTOS can be translated to VHDL, we can identify the reasonable subset of LOTOS to be translated to VHDL. The following subset is enough to describe hardware devices.

- **Data types:** The values used in LOTOS specifications will be restricted to the values implementable using VHDL. The set of usable values depends on the available VHDL tools. We will provide a LOTOS library with the allowed data types. That library defines the data types and data values that can be used in combination with a VHDL simulator or synthesizer.

- **Dynamic creation of processes:** LOTOS allows the dynamic creation of processes. It does not seem reasonable to use dynamically changing architectures when hardware is defined. Recursion among static processes is supported, providing that the static architecture is not changed.
- **Value negotiation:** Value negotiation is not only very powerful, but also very complex. In real situations it is possible to avoid it. This is not a restrictive limitation when modelling hardware, where value negotiation does not seem natural, and value passing is used instead. The value may be rejected by the receiver if it does not conform to the selection predicate.
- **Rendezvous:** After applying the above restrictions we conclude that multi-way rendezvous need not be implemented. Using the value passing convention, one of the partners puts out one or more data values and the other partners must accept them. Then the synchronization is reduced to: 1 to n ($n \geq 1$) value passing.
- **External interface:** Problems with the external interface mainly occur in relation to synchronization with the environment. As we described before, VHDL interprets signals as putting or getting values on or from a wire. Since the LOTOS interpretation is much more complex, two possible modes of interpretation for translation can be defined:
- **Asynchronous mode:** If we intend to make a synchronous interpretation of the LOTOS specification, the environment must be aware of this. The environment must follow some rules in order to provide a consistent interface to the VHDL generated code. These rules will require the use of more than one wire for each LOTOS gate, and the implementation of some kind of protocol.
- **Synchronous mode:** If interpretation is directly to the VHDL semantic model, the interface will demand one wire for each LOTOS gate, and communication will be the same as for any other VHDL piece of code. This mode of interpretation will require much more refined LOTOS specifications.

In our translation we deal with the asynchronous mode.

Following this model, we have implemented a tool called HARPO that translates LOTOS to VHDL automatically (see also [6], `http://www.dit.upm.es/~format/`). It has been tested by the telecommunications companies Italtel and Telefónica I+D. A further application found for HARPO is the automatic generation of VHDL test benches.

6.4. CONCLUSION

In this paper, we have related the abstract design description language LOTOS to the hardware description language VHDL. This approach can be used as a means to connect formal abstract design to lower levels of description, in the same way as VHDL has served as an input to high-level synthesis.

We do not know about other works translating LOTOS to VHDL, but there have been several approaches to translating from different specification languages to VHDL. Some of them also use a formal specification language. For example, in [20] systems are specified in Esterel and then translated to VHDL. In [12], the formal description technique Estelle is translated to VHDL in order to obtain correct descriptions automatically or semi-automatically. Furthermore, ExpressV-HDL translates StateCharts to VHDL. See [15] for a comparison of the result of this commercially available tool with the HARPO tool.

Less related are the works in which object orientation (OO) and formal description techniques (FDTs) are combined, like the INSYDE Project (integrated methods for evolving system design). At the analysis phase OMT (the OO methodology of Rumbaugh) is used to capture the functional requirements of the system, resulting in a conceptual model of it. The software parts of the design are then translated into SDL semi-automatically, while the hardware parts are mapped onto VHDL.

6.4.1. Acknowledgements

We would like to thank Natividad Martínez Madrid, Peter T. Breuer and Luis Sánchez Fernández for fruitful discussions and comments on the presentation of this work and on the example of the Ethernet Bridge. This work has been partially sponsored by ESPRIT project No. 6128 FORMAT (Formal Methods for Hardware Verification) and CICYT (Spanish Research Programme) project INTEGRAL.

REFERENCES

[1] R. Airiau, J. M. Bergé, and V. Olive. *Circuit Synthesis with VHDL*. Kluwer, 1994.

[2] P. J. Ashenden. *The VHDL Cookbook*. University of Adelaide, South Australia.

[3] J. M. Bergé, A. Fonkoua, S. Maginot, and J. Rouillard. *VHDL '92, The New Features of the VHDL Hardware Description Language*. Kluwer, 1993.

[4] T. Bolognesi and E. Brinksma. Introduction to the ISO Specification Language LOTOS. *Special Issue of "Computer Networks and ISDN Systems" on: Protocol Specification and Testing*, 14(1):25–59, 1987.

[5] C. Delgado Kloos and P. T. Breuer, editors. *Formal Semantics for VHDL*. Kluwer, 1995.

[6] C. Delgado Kloos, T. de Miguel, T. Robles, G. Rabay, and A. Marín. VHDL Generation from a timed extension of the formal description technique LOTOS within the FORMAT project. In F. Vajda and B. Graff Mortensen, editors, *Open System Design: Hardware, Software and Applications*, volume 38: 1–5, pages 589–596. Euromicro, North-Holland, 1993. Proceedings Euromicro'93, Barcelona, 6–9 September, 1993.

[7] H. Ehrig, W. Fey, and H. Hansen. *ACT ONE: An Algebraic Language with two Levels of Semantics*. Technical Report Bericht Nr. 83.103, Tech. Universität Berlin, 1983.

[8] R. E. Harr and A. G. Stanculescu. *Applications of VHDL to Circuit Design*. Kluwer, 1991.

[9] C. A. R. Hoare. *Communicating Sequential Processes. Communications ACM*, 21(8):666–677, 1978.

[10] C. A. R. Hoare. *Communicating Sequential Processes. Prentice Hall Int'l*, 1985.

[11] IEEE. *IEEE Standard VHDL: Language Reference Manual*. IS Std 1076–1993, The Institute of Electrical and Electronics Engineers, 1994.

[12] J.Wytrebowicz and S.Budkowski. *Communication Protocols Implemented in Hardware: VHDL Generation from Estelle*. Technical report, 1994. Proceedings VHDL-Forum for CAD in Europe, Sepetember 1994.

[13] L. Logrippo, M. Faci, and M. Haj-Hussein. *An Introduction to LOTOS: Learning by Examples*. Technical report, University of Ottawa, Dept. of Computer Science, Protocols Research Group, 1992.

[14] J. A. Mañas, J. Salvachúa, and T. de Miguel. *The TOPO Implementation of the LOTOS Multiway Rendezvous*. Technical report, Dpt. Telematics Engineering, Technical Univ. Madrid, Ciudad Universitaria, E-28040 Madrid, Spain, Jan. 1991.

[15] N. Martínez Madrid, A. Marín López, S. Deprés Díaz, C. Delgado Kloos and L. Sánchez Fernández. *Cremalleras y espirales: Como co-diseñar mejor*. In I Jornadas de Informática, 1995.

[16] R. Milner. *A Calculus of Communicating Systems*. LNCS. Springer-Verlag, Berlin, 1980.

[17] R. Milner. *Communication and Concurrency*. Prentice Hall, 1989.

[18] D. E. Ott and T. J. Wilderotter. *A designer's guide to VHDL synthesis*. Kluwer, 1994.

[19] J. Quemada, A. Azcorra, and D. Frutos. *TIC – A Timed Calculus for LOTOS.* In S. T. Vuong, editor, Formal Description Techniques, II, pages 195–209, Vancouver (CA), 1990. IFIP, Elsevier Science B.V. (North-Holland). Proceedings FORTE'89, 5–8 December, 1989.

[20] B. Raphael and T. Holzer. *From Esterel to VHDL: A Way to Write Proved VHDL.* 1993. Proceedings VHDL-Forum for CAD in Europe, March 1993.

6.5. APPENDIX: A COMPLETE EXAMPLE

This example is the specification of an Ethernet Bridge. The informal specification of the Bridge is the following: *An Ethernet Bridge operates below the Medium Access Control (MAC) Service Boundary, and is transparent to protocols operating above this boundary in the Logic Link Control (LLC). It interconnects separate Ethernet LANs (ISO-IEC 8802-3) that comprise a Bridged Local Area Network by relaying frames between the separate MACs of the bridged LANs. The basic functions performed by Bridges are frame forwarding, learning stations addresses and resolving loops in the topology with the spawning tree algorithm.*

We have not implemented the third functionality for simplicity. Incoming frames addresses are stored in the forwarding database. Frame forwarding is done by examining incoming frames and looking for their destination in the forwarding database. If this information is not found, the frame is broadcasted to the rest of the ports. Bridge learning means that when incoming frames are examined fo forwarding, the source address and the port identifier are updated in the database. Additional buffering should be provided for both inputs and outputs.

Here follows the LOTOS specification. It shows the parallel composition of six processes: Memory, Control, Concentrator and three different instantiations of Port. The composition uses the parallel explicit operator, which stands for synchronization in the listed gates and interleaved behaviour in the rest.

The definition of the data types is not included, but the operations are commented on the code. Basically, we define type frame, type frame_queue, type bridge_info, which adds a port identifier to a frame, and finally types table_info and table for storing addresses and their associated ports. The operation noport stands for address not in the table, and implies flooding of the frame.

```
specification
  eth_brdge [in1,in2,in3,out1,out2,out3]: noexit
library
  Boolean,NaturalNumber,Port,Frame,
        Frame_list,Bridge_info,Table,Table_info
endlib
behaviour
 HIDE req,ans,req_in,ans_out,income1,income2,income3,outcome IN
(Memory[req,ans](<> of table)
(* stores the table of addresses and ports *)
 |[req,ans]|
(* parallel synchronization in gates req and ans *)
  (Control[req,ans,req_in,ans_out]|[req_in,ans_out]|
    (Concentrator[income1,income2,income3,outcome,req_in,ans_out]
     |[income1,income2,income3,outcome]|
(* parallel synchronization in gates income and outcome *)
      (Port[in1,out1,income1,outcome]
(1,<> of frame_queue,<> of frame_queue)
       |[outcome]|
        Port[in2,out2,income2,outcome]
(2,<> of frame_queue,<> of frame_queue)
       |[outcome]|
        Port[in3,out3,income3,outcome]
(3,<> of frame_queue,<> of frame_queue)
(* the three ports are interleaved in gate incomeX with process
control and synchronized in gate outcome *)
))))
where
process Memory[addr_in,addr_out](Addr_DB: table)
: noexit := addr_in? tinf: table_info;
  (* Request of port *)
  addr_out! Lookup(Get_Dest(tinf),Addr_DB);
  (* Search in the table *)
  Memory[addr_in,addr_out](Update(tinf,Addr_DB))
  (* Update table *)
endproc
process Control [req,ans,a,b]: noexit := a? brinf: bridge_info;
  (* frame income from any of the ports *)
  req!Table_info(Get_Port(brinf), Get_Source(Get_Frame(brinf)),
      Get_Dest(Get_Frame(brinf)));
  (* Ask the memory for the destination port *)
  ans? outp : nat;
  (* Get the answer *)
  ([(outp eq noport)or(outp ne Get_Port(brinf))]
   ->b! Bridge_Info(outp,Get_Frame(brinf));
     Control [req,ans,a,b]
     (* Send the frame with the destination port to all the ports *)
 []
   [outp eq Get_Port(brinf)] ->
     Control [req,ans,a,b]
     (* Frame source and destination in the same ethernet *)
  )
endproc
process Concentrator
[in1,in2,in3,ouc,req_in,ans_out] : noexit:= in1? brinf: bridge_info;
   req_in! brinf;
   ans_out? brinfout: bridge_info;
   ouc! brinfout;
   Concentrator[in1,in2,in3,ouc,req_in,ans_out]
 []
```

```
    in2? brinf: bridge_info;
    req_in! brinf;
    ans_out? brinfout: bridge_info;
    ouc! brinfout;
    Concentrator[in1,in2,in3,ouc,req_in,ans_out]
  []
    in3? brinf: bridge_info;
    req_in! brinf;
    ans_out? brinfout: bridge_info;
    ouc! brinfout;
    Concentrator[in1,in2,in3,ouc,req_in,ans_out]
endproc
process Port[inp,outp,inc,ouc](id: nat, flistin,flistout:
      frame_queue): noexit := inp? frin: frame;
    Port[inp,outp,inc,ouc](id,Append(frin,flistin),flistout)
    (* Receive frame from ethernet number id, store it in flistin *)
  []
    ouc? frout: bridge_info;
    (
     [(Get_Port(frout) eq noport)or (Get_Port(frout) eq id)]->
      Port[inp,outp,inc,ouc](id,flistin,
           Append(Get_Frame(frout),flistout))
      (* Receive frame from bridge, store it in flistout *)
  []
     [Get_Port(frameout) ne id]->
      Port[inp,outp,inc,ouc](id,flistin,flistout)
      (* Receive frame from bridge; frame is not for this ethernet *)
    )
  []
    [not(IsEmpty(flistout))]->
    (outp! First(flistout);
    Port[inp,outp,inc,ouc](id,flistin,Extract(flistout)))
     (* outp to ethernet id the first frame of the flistout queue *)
  []
    [not(IsEmpty(flistin))]->
     (inc! Bridge_info(id,First(flistin));
    Port[inp,outp,inc,ouc](id,Extract(flistin), flistout))
     (* outp to bridge the first frame of the flistin queue *)
endproc
endspec
```

We can automatically translate this specification to VHDL with our tool HARPO, and we obtain five entities (Port, Concentrator, Control, Memory, and eth_brdge). The last one contains the information relating to the interconnection of the others, and the values with which they are first instantiated. It uses two components synch_op and interl_op, which stand for full synchronization or interleaving. The rest of the entities are FSMs described in VHDL. Let us have a look at the combinational part of one of them, for example the ENTITY PORT:

```
COM: PROCESS(clk)
--  Auxiliary variable definitions deleted
VARIABLE vVal_inc:      std_ulogic_vector (size_bridge_info downto 0);
VARIABLE vVal_outp:     std_ulogic_vector (size_frame downto 0);
VARIABLE uflistin:      std_ulogic_vector (size_frame_queue downto 0);
VARIABLE uflistout:     std_ulogic_vector (size_frame_queue downto 0);
VARIABLE ufrin:         std_ulogic_vector (size_frame downto 0);
VARIABLE ufrout:        std_ulogic_vector (size_bridge_info downto 0);
```

```
BEGIN
-- Variable assignments deleted
IF CS = '1' and clk = '1' THEN
CASE state IS
WHEN 0 => next_state <=1;
          uid:= sid;
          uflistin:= sflistin;
          uflistout:= sflistout;
WHEN 1 => next_state<=2;
          vReset_inp:= '0';
          IF not(IsEmpty(uflistout)) THEN
            vReset_outp:= '0';
            vVal_outp:= First(uflistout);
          END IF;
          IF not(IsEmpty(uflistin)) THEN
            vReset_inc:= '0';
            vVal_inc:= bridge_info(uid,First(uflistin));
          END IF;
          vReset_ouc:= '0';
WHEN 2 =>IF Ack_inp='1' THEN
            vOk_inp:= '1';
            next_state<=3;
          END IF;
          IF Ack_outp='1' THEN
            vOk_outp:= '1';
            IF not(IsEmpty(uflistout)) THEN
              next_state<=4;
            ELSE  vOk_outp:= '0';
            END IF;
          END IF;
          IF Ack_inc='1' THEN
            vOk_inc:= '1';
            IF not(IsEmpty(uflistin)) THEN
              next_state<=5;
            ELSE  vOk_inc:= '0';
            END IF;
          END IF;
          IF Ack_ouc='1' THEN
             vOk_ouc:= '1';
             next_state<=6;
          END IF;
WHEN 3 => IF Ack_inp='1' THEN
             vOk_inp:= '1';
             ufrin:=inp;
             next_state<=7;
          ELSIF Nack_inp='1' THEN
             next_state<=1;
          END IF;
WHEN 6 => IF Ack_ouc='1' THEN
             vOk_ouc:= '1';
             ufrout:=ouc;
             next_state<=8;
          ELSIF Nack_ouc='1' THEN
             next_state<=1;
          END IF;
WHEN 7 => next_state <= 1;
          uflistin:= Append(ufrin, uflistin);
```

```
WHEN 8 => IF(Get_Port(ufrout) eq natToStdUlogicVector(noport)
            or Get_Port(ufrout) eq uid) THEN
               next_state <= 9;
          ELSIF Get_Port(ufrout) ne uid THEN
               next_state <= 1;
          END IF;
WHEN 9 => next_state <= 1;
          uflistout:= Append(Get_Frame(ufrout),uflistout);
WHEN 4 => next_state <= 1;
          uflistout:= Extract(uflistout);
WHEN 5 => next_state <= 1;
          uflistin:= Extract(uflistin);
END CASE;
END IF;
 last_state <= vlast_state;
 Reset_ouc <= vReset_ouc;
 Ok_ouc <= vOk_ouc;
 Reset_inc <= vReset_inc;
 Ok_inc <= vOk_inc;
 Val_inc <= vVal_inc;
 Reset_outp <= vReset_outp;
 Ok_outp <= vOk_outp;
 Val_outp <= vVal_outp;
 Reset_inp <= vReset_inp;
 Ok_inp <= vOk_inp;
END PROCESS;
```

7

USING AN X-MACHINE TO MODEL A VIDEO CASSETTE RECORDER

Matt Fairtlough, Mike Holcombe, Florentin Ipate, Camilla Jordan, Gilbert Laycock and Duan Zhenhua

Formal Methods and Software Engineering Research Group, (Formsoft), Department of Computer Science, University of Sheffield, U.K.

ABSTRACT

The problem of constructing a simple formal specification of a dynamic, time dependent system is addressed here. We use an intuitive state machine model which can be developed in a series of stages, each successive refinement adding new features and addressing new issues related to the design of the specification. The model used is fully general, unlike traditional state machine models, and can be supported by a test generation method that will provide a basis for an integrated design for the test specification method. The case study is the specification of a video cassette recorder system which is defined in a formal way with the minimum of mathematical notation.

7.1. INTRODUCTION

A number of formal specification languages and methods have been proposed in recent years and applied to a number of case studies. They range from model based languages, such as Z [1] and VDM [2], to executable algebraic languages like OBJ [3] and process algebra approaches, principally CCS [4] and CSP [5].

The use of temporal and other formal logics for the description and verification of systems is also of research interest. These methods all have their advantages and disadvantages or limitations. In many cases they do not, in their original form, address the problem of modelling time in real-time systems (even temporal logic, as originally formulated, does not describe explicit time intervals directly). Coping with time has usually required the augmentation of the language by special constructors and operations which have often caused a significant increase in the complexity of the notation and methods and a consequent reduction in its attractiveness as a useful tool for designers in industry to use. The intense symbolic and overt mathematical appearance of the notations has also impeded their widespread use. Let us consider, for example, Milner's CCS (calculus of communicating systems). This calculus has been used to model a number of complex systems, electronic and otherwise, and clearly could be used to construct a formal model of a video cassette recorder (VCR), following similar lines to this paper, provided the full value-passing version is used. CCS was devised to give a general theoretical account of concurrent, asynchronous, non-deterministic computation and is most useful for analysing the communication structure of a system; indeed, there is a highly automated tool (the Concurrency Workbench) for doing this for finite-state models. It is more difficult to use CCS to model time or the data that flow in a typical system, whereas in our approach considerable emphasis is placed on analysing data and the functions which process it, and time can be modelled in a straightforward way.

Here we are concerned with deterministic systems without many parallel components, and this restriction will allow the straightforward development of simulation and testing tools which take data and the flow of time into account. In contrast, and despite the age and pedigree of CCS, tools which support the full value-passing version are only now being developed and still face theoretical problems, while the issue of testing has hardly been addressed.

There have been a number of approaches which combine the intuitive attractive idea of a state machine with various techniques for incorporating time. None of these are fully general in the sense that any computational process can be modelled in this way.

We wish to propose an alternative, but related, approach based on a similar graphical approach which is precise, fully general and formal and lies at the heart of computational modelling. It is a blend of state diagrams and simple formal descriptions of data-types and functions which can easily be expressed in a language such as Z or as functions in ML [6] or similar functional languages.

The theory of finite state machines is a popular approach to the theory of computation, it is widely used in a number of key technologies such as sequential hardware design, compiler technology and real-time systems design. It does, however, suffer from a limitation in its computational power and cannot be used effectively in large scale problems. A way of managing this problem has been proposed by Harel, namely the Statechart approach [7], and this is undoubtedly an improvement. However, the model can still become very cumbersome and, in some examples we have explored, produces state spaces significantly larger than the method proposed here. We demonstrate the use of a generalisation of the finite state machine method as a specification and verification mechanism which is:

- rigorous;
- descriptive;
- models data and date processing explicitly;
- handles time simply;
- fully general;
- not too general (in the sense that it only specifies computationally possible algorithms provided the basic functions involved are computable; see [8]);
- is easy to use;
- can be used effectively for design refinement;
- can be executed with appropriate tools;
- is the basis for a formal test set generation method.

The model we have chosen, both as a basis for theoretical work in the theory of computability and the theory of testing and as a basis for a formal specification language, is the X-machine. Introduced by Eilenberg in 1974 [9], it has received little further study. Holcombe [10] proposed the model as a possible specification language, and since then a number of investigations have demonstrated that this idea is of great potential value to software engineers. In its essence an X-machine is like a finite state machine but with one important difference. A basic data set, X, is identified together with a set of basic processing functions, Φ, which operate on X. Each arrow in the finite state machine diagram is then labelled by a function from Φ, the sequences of state transitions in the machine determine the processing of the data set and thus the function computed. The data set X can contain information about the internal memory of a system as well as different sorts of output behaviour, so it is possible to model very general systems in a transparent way. It is best to separate the control state of the system from the data state since this allows much more scope for organising the model to ensure a small and manageable state space. This is done easily with this method; the set X is often an array consisting of fields that define internal structures such as registers, stacks, database filestores, input information from various devices, models of screen displays and other output mechanisms. The functions will read inputs, datafiles, internal memory, write to all of these, refresh displays, etc.

This paper examines the method through a simple case study involving the specification and verification of a simple VCR machine. The method demonstrates three basic types of specification refinement/extension which are very natural and which can easily be described in a formal setting, if desired. We first examine the basic model of the system demonstrating the main features of the X-machine method. This involves the identification of a number of basic states of the VCR controller together with the description of the data space and functions that the computation is based upon. We then show how the state space can be enlarged to deal with increased functionality in a simple and intuitive way. The next two sections introduce the problem of setting a time period and integrating a simple clock into the model. Following this we model the tape and its contents to form a complete, if simple model. We then consider how a simple animation can be derived from the model, and we end with some consideration of the verification of different types of properties of the model. The method is evaluated in the light of the experience in developing this model and of the needs of a user friendly formal method that can be used to describe and reason about dynamic, real-time systems.

7.2. THE BASIC MODEL

Those X-machines in which the input and the output sets behave as orderly streams of symbols are called *stream* X-machines. The basic idea is that the machine has some internal memory, M, and the stream of inputs determine, depending on the current state of control and the current state of the memory, the next control state, the next memory state and any output value.

So if Σ is the set of possible inputs and Γ represents the set of possible outputs we put

$X = \text{seq } \Gamma \times \text{M} \times \text{seq } \Sigma$

Each processing function $\phi : \text{seq } \Gamma \times \text{M x seq } \Sigma \rightarrow \text{seq } \Gamma \times \text{M} \times \text{seq } \Sigma$ is of the form whereby, given a value of the memory and an input value, ϕ can change the memory value and produce an output value, the input value is then discarded. There is an initial state, and all states are terminals.

Our initial aim was to produce a very simple but formal model showing the basic features of a VCR.

A *Stream X-machine* (SXM) consists of a 6-tuple,

$M = (X, Q, \Phi, F, I, T)$

In this tuple $X = \text{seq } \Gamma \times \text{M} \times \text{ seq } \Sigma$ consists of the stream of outputs, the memory, and the stream of inputs. In our first model we called this X_1. The inputs are the pressing of named buttons, and the outputs are the symbols displayed by the machine to show whether it is on, which state it is in (OP) and which channel is set (Chan). Since, under some circumstances, there is no display, we have included the symbol $\perp$ to indicate this. In this first version the memory simply consists of a number representing the channel.

$X_1 = \text{seq } \Gamma_1 \times M_1 \times \text{seq } \Sigma_1$
$\Sigma_1 = \{\text{on, off, play, record, ch\_up, ch\_down, stop}\}$
$\Gamma_1 = \text{B} \times OP_1 \times \text{Chan}_\perp$
$OP_1 = \{\text{Idle, Play, Record, Off}\}$
$\text{Chan} = \{1..\text{k}\}$
$M_1 = \text{Chan}$
$\mathbf{B = \{0, 1\}}$

Q_1 is the set of states, $Q_1 = \{\textit{Off, Idling, Playing, Recording}\}$, and F_1 is the next state function. These can be seen in the state-transition diagram, Figure 1. Φ_1 is the set of processing functions,

Φ_1 = {sw-on, sw-off, play, record, channel, stop}$\cup${continue$_q$ | $q \in Q_1$}

I and T are the sets of initial and terminal states respectively. Here $I_1 = \{Off\}$, $T_1 = Q_1$ and $m_1 = 1$ (the initial state of the memory).

A *stream X-machine* like this has the property that there is precisely one output value for each input value. The best way to describe an X-machine is to produce the state-transition diagram and details of the processing functions. This first machine is a stream X-machine. It is intended to be deterministic and complete with respect to seq Σ , see [11], that is, it is well defined for any sequence of inputs.

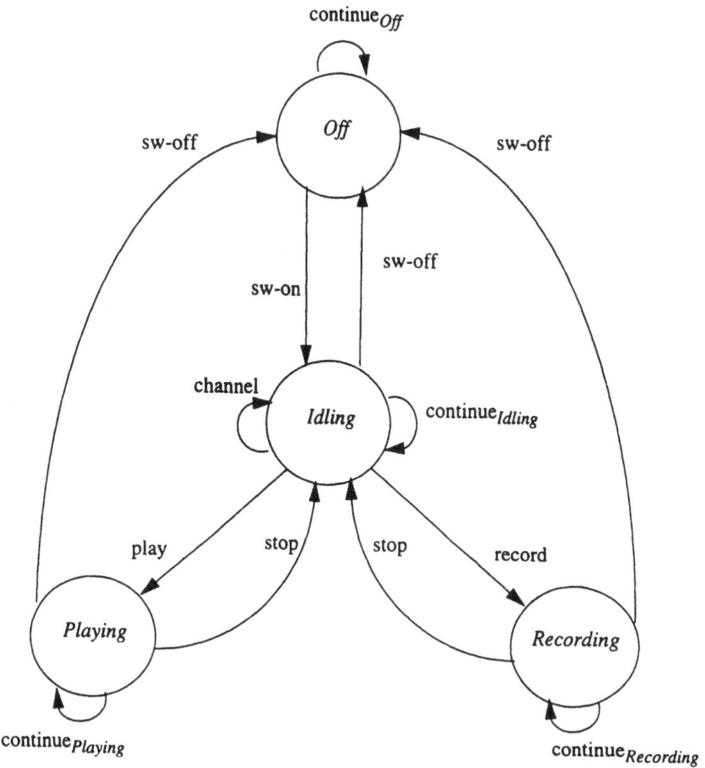

Figure 1: An X-machine State Diagram for the Basic Model

7.2.1. The Processing Functions

In order to describe the machine properly we need to give details of the processing functions. The argument of a processing function is a member of X and so can be regarded as a triple. The result is also a member of X. To avoid unnecessary detail, we assume that if the argument is not of the form given, then the result is undefined. To cope with channel changes, we use the unary operators *Ch_up* and *Ch_down* which are defined as:

$Ch\_up\,(x) = x + 1$ if $0 < x < \mathrm{k}$, else 1
$Ch\_down\,(x) = x - 1$ if $1 < x < \mathrm{k}$, else k

The processing functions, defined next, have the effect of removing the current input symbol, altering the memory and placing an output symbol on the end of the output string. Thus, if

$G \in \mathrm{seq}\Gamma_1, \mathrm{c} \in \mathrm{M}_1, \sigma \in \Sigma_1, S \in \mathrm{seq}\,\Sigma_1$

then

$\phi(G, c, \sigma :: S) = (\, G :: \gamma, c', S\,)$ for suitable $\gamma \in \Gamma, c' \in M_1$.

sw-on(G, c, on :: S)	=	(G :: (1, Idle, c), c, S)
sw-off(G, c, off :: S)	=	(G :: (0, Off, $\perp_{chan}$), c, S)
channel(G, c, ch_up :: S)	=	(G :: (1, Idle, *Ch_up*(c)), *Ch_up*(c), S)
channel(G, c, ch_down :: S)	=	(G :: (1, Idle, *Ch_down*(c)), *Ch_down*(c), S)
play (G, c, play :: S)	=	(G :: (1, Play, c), c, S)
record (G, c, record :: S)	=	(G :: (1, Record, c), c, S)
stop (G, c, stop :: S)	=	(G :: (1, Idle, c), c, S)
continue$_{Off}$(G, c, s :: S)	=	if $s \in$ {play, record, off, ch_up, ch_down, stop} then (G :: (0, Off,$\perp_{chan}$), c, S)
continue$_{Idling}$ (G, c, on :: S)	=	(G :: (1, Idle, c), c, S)
continue$_{Idling}$ (G, c, stop :: S)	=	(G :: (1, Idle, c), c, S)
continue$_{Playing}$ (G, c, s :: S)	=	if $s \in$ {play, record, off, ch_up, ch_down, stop} then (G :: (1, Play, c), c, S)
continue$_{Recording}$ (G, c, s :: S)	=	if $s \in$ {play, record, off, ch_up, ch_down, stop} then (G :: (1, Record, c), c, S)

7.3. EXTENDING THE MACHINE TO HANDLE FAST-FORWARD AND REWIND OPERATIONS

Having got our basic machine formalised, it is easy to add additional features like fast-forward and rewind, particularly as the machine has a hub format with *Idling* as the hub. We simply add two new states and some appropriate processing functions. The underlying data type (memory) is unchanged. This is the essence of the first type of specification refinement which is very easy to do as long as some care is taken.

7.3.1. The Tuple

We only include the changed members.

Q_2 = {*Off*, *Idling*, *Playing*, *Recording*, *FFWDing*, *RWDing*}
Σ_2 = {on, off, play, record, ch_up, ch_down, stop, ffwd, rwd}
Φ_2 = {sw-on, sw-off, play, record, channel, stop, fast_forward, rewind}$\cup$$\{\text{continue}_q \mid q \in Q_2\}$
I_2 = {*Off*, *Idling*}
OP_1 = {Idle, Play, Record, FF, RWD, Off}

F_2 is shown in Figure 2.

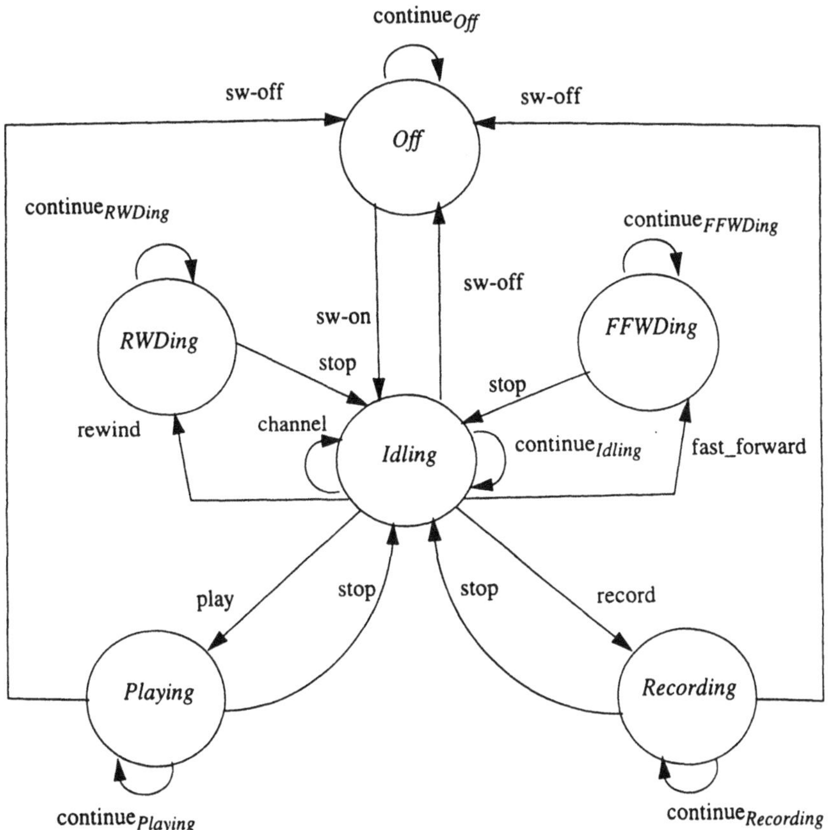

Figure 2: *X*-Machine State Diagram for the Extended Machine (Including Fast Forward and Rewind)

7.3.2. The Processing Functions

fast_forward(G, c, ffwd :: S) = (G :: (1, FF, c), c, S)

rewind(G, c, rwd :: S) = (G :: (1, RWD, c), c, S)

continue$_{Playing}$(G, c, s :: S) = if $s \in$ {play, record, off,ch_up, ch_down, ffwd, rwd} then (G :: (1, Play, c), c, S)

continue$_{Recording}$(G, c, s :: S) = if $s \in$ {play, record, off, ch_up, ch_down, ffwd, rwd} then (G :: (1, Play, c), c, S)

continue$_{FFWDing}$(G, c, s :: S) = if $s \in$ {play, record, off, ch_up, ch_down, ffwd, rwd} then (G :: (1, FF, c), c, S)

continue$_{RWDing}$(G, c, s :: S) = if $s \in$ { play, record, off, ch_up, ch_down, ffwd, rwd } then (G :: (1, RWD, c), c, S)

7.4. PROGRAMMING THE VCR

The previous sections show how a basic model of the behaviour of a VCR can be built and extended using an *X*-machine. However, suppose we want to model a new aspect of the VCR's behaviour as a completely separate machine (with the intention of combining them at a later stage). This modular approach allows us to concentrate on the most important aspects of each part of the machine without being overwhelmed by components that are not relevant.

As an example, we will model the programmed control aspect of a VCR as a new *X*-machine.

The actual VCR implementation will need to have some form of clock, which is continually updated.

The set of states is

Q_3 = {*Programming*, *ProgWaiting*, *Recording*}

with *Programming* as the only initial and terminal state.

Each actual input is either a tick, corresponding to the passage of time, or a user input which occurs concurrently with a tick. In other words, the input set is as follows:

$\Sigma_3 = (\mathbf{T} \times \mathbf{T}) \cup \{\text{set, unset, tick}\}$

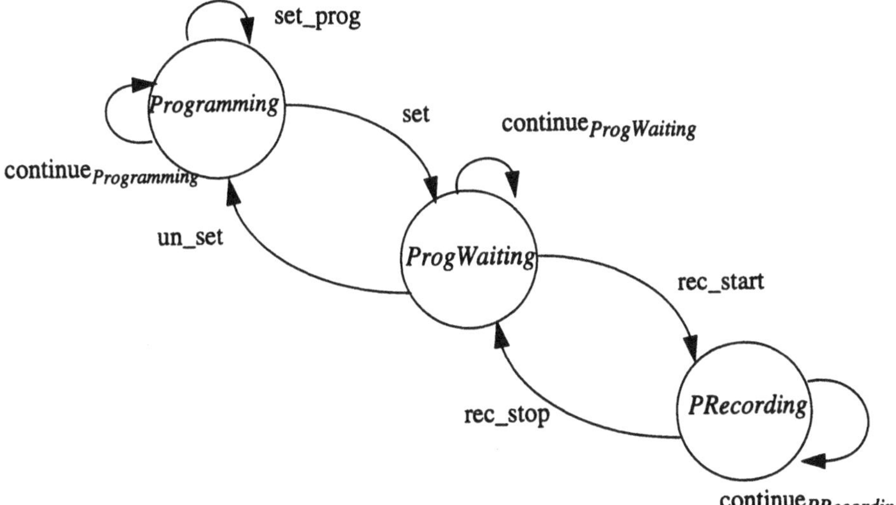

Figure 3: State Transition Diagram for Programmable Recording

T is a type suitable for recording time. For instance, **T** could simply be the natural numbers, starting from some arbitrary point and advancing by one for every hundredth of a second. Alternatively, it could be a complex data type with fields for the year, month, day, hour, minute and second. Critical factors are that we have to be able to compare elements and discover whether they are equal, or which element is larger and to be able to increment time values by one unit.

The memory is a little more complex,

$M_3 = \mathbf{T} \times (\mathbf{T} \times \mathbf{T})$

It consists of two components. The first records the current time. The second component is a pair of **T** elements, corresponding to the program settings: a program start time and a program end time.

The processing functions are as follows:

setprogram$(G, (t, (x, y)), (st, en) :: S) = (G :: newg, (t + 1, st, en), S)$

set$(G, (t, (\mathrm{st}, \mathrm{en})), \mathrm{set} :: S) = (G :: newg, (t + 1, st, en), S)$

continue$_{Programming}(G, (t, (st, en)), s :: S) =$ if $s \neq$ set and $s \neq (st, en)$ then $(G :: newg, (t + 1, st, en), S)$

rec_start$(G, (t, (st, en)), \mathrm{tick} :: S) =$ if $t + 1 = st$ then $(G :: newg, (t + 1, st, en), S)$

un_set$(G, (t, (st, en)), \mathrm{unset} :: S) = (G :: newg, (t + 1, st, en), S)$

continue$_{ProgWaiting}(G, (t, (st, en)), s :: S) =$ if $t \neq st - 1$ then $(G :: newg, (t + 1, st, en), S)$

stop$(G, (t, (x, y)), (st, en), \mathrm{tick} :: S) =$ if $t + 1 = en$ then $(G :: newg, (t + 1, st, en), S)$

$\text{continue}_{PRecording}(G, (t, (x, y)), (st, en), s :: S) = \text{if } (t > st) \text{ and } (t < en - 1) \text{ then}$
$(G :: newg, (t + 1, st, en), S)$

where *newg* is the new output and $(x, y) \in$ T.

These processing function definitions can be explained informally:

setprogram is the processing function executed when the user enters a new programme for the machine. In other words, there is a pair (*st*, *en*) on the input sequence corresponding to the start-time and end-time that the user wants to set. For simplicity, we assume that the user will only enter programs with $st < en$ and it would be a simple matter to extend the machine to check this. Also, the user could enter a program with a start- and end- time earlier than the current time. In this case the programmed recording will never start. Once again, it would be a simple matter to extend the machine to check this.

set is the processing function that is applied when the user presses the set input, putting the VCR into a state where it waits for the programmed start time to arrive. If the set input occurs after the programmed start time has already passed, then the VCR will never start recording.

$continue_{programming}$ is the processing function applied while the machine is idle in the *Programming* state, waiting for the user to enter a new programme, or to press the setI input. The only effect it has on the machines memory is to advance the current time by one unit.

rec_start is the processing function that starts a programmed recording. It will only be applied if the current time (*t*) is exactly one instant before the start time (*st*) of the programme. So, if the start time has already passed before the machine moves to the *ProgWaiting* state, then it will never start recording.

un_set is the processing function applied when the user presses the unset input and returns to the state where a new program can be entered.

$continue_{ProgWaiting}$ is the processing function applied while the machine is idly waiting in the *ProgWaiting* state. It can only be applied if the current time is before or after the start time for the programmed recording.

stop is the processing function that is applied when programmed recording is completed. It can only be applied when the current time (*t*) is exactly one instant before the programmed recording end time (*en*).

$continue_{PRecording}$ is the processing function applied repeatedly during the programmed recording. It simply advances the current time by one instant and is only defined if the current time (*t*) is between the programmed start time (*st*) and the programmed end time (*en*).

7.5. COMBINING THE PARTS OF THE MODEL

7.5.1. Combined Model

If the tape is ignored, a complete model of the VCR is obtained by linking the sub-machines from Sections 7.3 and 7.4, respectively. We show first how this is done for this particular case. We will then describe a method of linking which works in general. This demonstrates another approach to refinement. In order to link the machines, we will need new inputs which have the role of transferring the action from one machine to the other.

The two extra inputs used are setprogI and unsetprogI. The set of states, the input alphabet

and the memory of the resulting machine are as follows.

$$Q = \{Off, Idling, Playing, Recording, FFWDing, RWDing\} \cup \{Programming, ProgWaiting, Recording\}$$

$$\Sigma = \{\text{on, off, play, record, ch\_up, ch\_down, stop}\} \cup (\mathbf{T} \times \mathbf{T}) \cup \{\text{set\_prog, unset\_prog}\}$$

$$M = \text{Chan} \times \mathbf{T} \times (\mathbf{T} \times \mathbf{T})$$

We are glossing over the output, but it includes a display of the current time and the current channel together with any current function (Play, Record, etc.).

There are two transitions linking the two sub-machines:

- one from *idling* to *Programming* labelled with set_prog,
- one from *Programming* to *idling* labelled with unset_prog.

Any function f in the first (time-free) machine is replaced by a function F in the resulting machine such that if

$$f(G, c, i :: S) = (G :: g, c', S)$$

then

$$F(G,(c, t, st, en), i :: S) = (G :: g, (c', t+1, st, en), S)$$

In fact, these functions do not change the memory of the second machine apart from updating the clock.

For a function, f, in the "set record" machine, we again use the originals in the obvious way, they do not affect the channel since we assume that the set record operations require the channel to be set in the *Idling* state.

Therefore, a function f from the initial (timer) machine is replaced by F such that if

$f(G, (t, st, en), i :: S) = (G:: g, (t + 1, st', en'), S)$

then

$F(G,(c, t, st, en), i :: S) = (G:: g, (c', t + 1, st', en'), S)$

The new transitions set_prog and unset_prog are defined as follows:

set_prog(G, (c, t, st, en), set_prog :: S) = (G :: γ_1 , (c, t, st, en), S)

unset_prog(G, (c, t, st, en), unset_prog :: S) = (G :: γ_2 , (c, t, st, en), S)

where γ_1 and γ_2 are two appropriate output characters.

To see if this makes sense, we can consider a stream of inputs and see what happens to M where m = (n, t, -, -) is the initial memory state.

The following stream of inputs is "applied":

[tick, tick, tick, on, tick, ch_up, tick, set, tick, , (st, en), tick, set_prog]

The new state of the memory is:

($Ch\_up(n)$, $t + 11$, st, en)

When a future tick causes $t + x$ to be equal to st, then the recorder starts, assuming $t + 11 < st$.

7.5.2. Hiding

In the next two sections we give a more formal description of the linking process. This would be needed in order to prove results about the linked machine.

The advantage of X-machines is that they can be displayed graphically. This advantage is lost, however, when the complications of the machine obscure the picture. The aim of hiding is to simplify the overall picture by hiding details. The principle approach when faced with the problem of combining two independently developed machines such as we have discussed above is to first collapse each machine into one which computes the same function with a single state and then to construct, in an appropriate way, a new machine from the two separate single state machines. Once we have achieved this, we can unpack the new machine to uncover the detail at the level of the original machines. This process is convenient and allows us to maintain the essential semantics of the situation.

We will assume, for the moment, that we have a stream X-machine denoted by $A = (X, Q, \Phi, F, I, T)$ where $X = \text{seq } \Gamma \times M \times \text{seq } \Sigma$. We can regard this as a function $A : M \times \text{seq } \Sigma \to \text{seq } \Gamma$. We can then define a new machine $A' = (X', \{A\}, \Phi', F', I', T')$ where $X' = \text{seq } \Gamma \times M' \times \text{seq } \Sigma$, which computes the same function A. The new machine will have a single state A and a single arc labelled ϕ'. The memory will be the Cartesian product of the old memory with the old set of states, i.e. $M' = M \times Q$. $\Phi' = \{\phi'\}$ will be defined as follows:

given

$\gamma \in \Gamma, G \in \text{seq } \Gamma, m, m' \in M, q, q' \in Q, s \in \Sigma, S \in \text{seq } \Sigma$

then

$\phi'((G, (m, q), s :: S)) = (G :: g, (m', q'), S)$

corresponds to

$F(q, \phi) = q'$ and $\phi((G, m, s :: S)) = (G :: g, m', S)$

We claim that A' will inherit the properties of determinism and completeness from A [14].

7.5.3. Linking

Suppose we have two single state machines A and B. We can form a new X machine A ◊ B by linking them as follows:

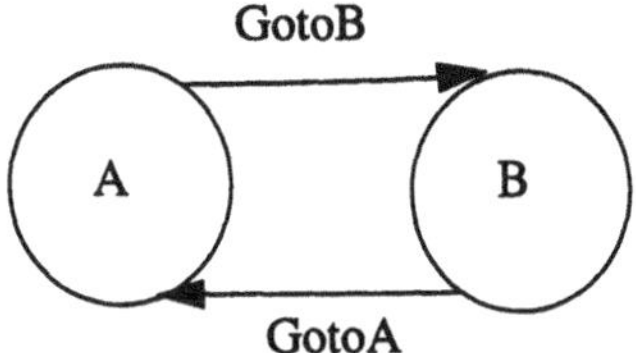

Figure 4: Machines A and B Combined into A.B

$A \quad = \quad \{X_A, \{A\}, \Phi_A, F_A\}$
where $X_A = \text{seq } \Gamma_A \times (M_A \times Q_A) \times \text{seq}\Sigma_A$,

$B \quad = \quad \{X_B, \{B\}, \Phi_B, F_B\}$
where $X_B = \text{seq}\,\Gamma_B \times (M_B \times Q_B) \times \text{seq}\Sigma_B,$

$A \lozenge B \quad = \quad \{X_{AB}, \{A, B\}, \Phi_{AB}, \mathrm{F}_{AB}\}$
where $X_{AB} = \text{seq}\,\Gamma_{AB} \times (M_{AB} \times Q_{AB}) \times \text{seq}\Sigma_{AB},$

$\Gamma_{AB} \quad = \quad (\Gamma_A \cup \Gamma_B \cup \{null\}) \times \{A, B\},$
$\Sigma_{AB} \quad = \quad \Sigma_A \cup \Sigma_B \cup \{GoA, GoB\},$
$\mathrm{M}_{AB} \quad = \quad (M_A \cup M_B) \times (Q_A \cup Q_B)$
$GotoA(G, (m, q), GoA :: S) = (G :: (null, A), (m', q'), S)$
$GotoB(G, (m, q), GoB :: S) = (G :: (null, B), (m', q'), S)$

This definition allows for hidden states to be kept in the memory. The notation (m', q') indicates that the functions *GotoA*, *GotoB* may modify the memory. The modification will depend on the particular example. Determinism and completeness properties will be preserved by this form of link.

In the model of the VCR we are investigating, we did not formally hide the two machines when linking them. Hiding is useful when developing more complex machines. It also simplifies the description of the general linked machine. For relatively simple models it may not be necessary to formally go through the process of hiding.

7.6. ADDING A MODEL OF THE VIDEO TAPE

We now consider a third type of refinement which involves the development of the memory without introducing any new states. The processing functions will be adapted to match the new memory structure. This, again, presents no real problems as long as we are consistent and careful.

7.6.1. The Data Type Tape.

Firstly, we need a data type for the tape which embodies all the properties of a video tape relevant to our specification:

- a sequence of frames, each one corresponding to one moment of TV, this can be captured by knowing the channel and time that the moment occured;
- the length of the tape;
- the current position of the tape, i.e. the frame number of the exposed part of the tape.

Tape $= (\mathrm{N} \rightarrow (\mathrm{Chan} \times \mathrm{T})) \times \mathrm{N} \times \mathrm{N}$

We can supply some useful functions for manipulating tapes:

TapePlay: **Tape** $\rightarrow$ **Tape** $\times (\mathrm{Chan} \times \mathrm{T})$
TapePlay $((tp, h, e)) \quad = \quad$ if $e < h$ then $((tp, h, e{+}1), tp(e))$

else if $h = e$ then $((tp, h, e), tp(e))$

TapeRecord: **Tape** × Chan → **Tape**

TapeRecord $((tp, h, e), (ch, t))$ = if $e < h$ then $((tp \oplus \{e \rightarrow (ch, t)\}), h, e{+}1)$

TapeFFD: **Tape** → **Tape**

TapeFFD $((tp, h, e))$ = if $e < h - 5$ then $(tp, h, e + 5)$
else if $h - 4 < e < h$ then $(tp, h, e + 1)$

TapeRWD: **Tape** → **Tape**

TapeRWD $((tp, h, e))$ = if $e > 5$ then $(tp, h, e - 5)$
else if $1 < e < 6$ then $(tp, h, e - 1)$
else if $1 = e$ then (tp, h, e)

TapePositionOK: **Tape** → **B**

TapePositionOK $((tp, h, e))$ = if $0 < e < h$ then true
else false

Here ⊕ is the function override construction and the fast forward and rewind functions move the tape through 5 positions each operation.

7.6.2. Integrating the Tape with the VCR

Given this type, we can add it to the *X*-machine model. No new states will be needed, but the data type (*X*) will have to be expanded to include a tape component, and every processing function will have to be extended. For the time being, we just add a tape to the machine without concerning ourselves over how to eject and change tapes.

7.6.2.1. A Tape in the Machine

The first step is to extend the internal memory of the machine to incorporate a tape:

$M' = M \times$ **Tape**

The only other change necessary to the internal data type is that the output data type needs to be extended to include a component for the current tape output:

$\Gamma' = \Gamma \times (\text{Chan} \times \text{T})$

Now all of the processing functions need extending. Rather than rewriting every single processing function, we can notice that there are only a few patterns to the necessary changes.

1. Changes to functions where the tape is irrelevant (such as switching the machine on, setting up a program and so on).

- Consider a processing function that has no effect on the tape, ϕ :
  $\phi(G, m, S) = (G', m', S')$
  then we can define the new version (ϕ') as follows:
  $\phi'(G, (m, tp), S) = (G', (m', tp), S')$

2. Changes to functions that require a change to the tape (such as playing the tape, recording, rewinding, and so on). These functions can be further subdivided according to the *way* they change the tape. For τ, $\tau' \in$ **Tape** and $sig \in$ Chan × T we have:

- By playing the tape.
  Consider a processing function, ϕ :
  $\phi(G, m, S) = (G :: g, m', S')$
  then the new version ϕ' is :
  $\phi'(G, (m, \tau), S) =$ if *TapePositionOK*(τ) then $(G :: (g, sig), (m', \tau'), S')$
  where *TapePlay*(τ) = (τ', sig)

- By recording on the tape.
  Consider a processing function, ϕ :
  $\phi(G, m, S) = (G :: g, (x, ch, t, y), S')$
  then the new version ϕ' is :
  $\phi'(G, (m, \tau), S) =$ if *TapePositionOK*(τ)
  then $(G :: (g, sig), ((x, ch, t, y), \tau'), S')$
  where *TapeRecord*(τ, sig) = (τ') and $sig = (ch, t)$

- By fast-forwarding or rewinding the tape.
  Consider a processing function, ϕ :
  $\phi(G, m, S) = (G :: g, m', S')$
  then the new version ϕ' is :
  $\phi'(G, (m, \tau), S) =$ if *TapePositionOK*(τ) then $(G :: (g, sig), (m', \tau'), S')$
  where *TapeFFD*(τ) = (τ') or *TapeRWD*(τ) = (τ')

7.7. ANIMATION AND EXECUTION OF THE SPECIFICATION

It is straightforward to represent a finite state machine in a functional language such as ML; once we have defined the states and arc-labels as enumerated types, the construction of the transition function from the transition diagram is entirely mechanical. Such a finite state machine underlies every SXM if we view the processing functions as simple arc-labels, and we can use this as the scaffolding for the SXM. If we now make the very reasonable assumption that the input alphabet, memory and transition functions of our SXM are computable (as is the case in the models we present here) then the underlying basis of our machine can be represented as an ML type and the processing functions as ML functions on that type. The resulting code follows the specification given in the previous sections very closely, which means that the code is likely to implement the specification correctly, and suggests that ML is a good choice of programming language for the animation of SXMs.

A prototype generic SXM animator has been constructed and tested on some modules defining SXMs such as the one described in this paper and works satisfactorily. Work is in progress to construct a graphical design tool for the design and animation of SXMs, which should greatly facilitate the correct construction and simulation of substantial designs.

7.8. VERIFICATION ISSUES

We will now consider how the final model satisfies some example user requirements, in other words how the formal specification may be verified against the informal user requirements.

User requirement UR1. The user wishes to record a programme which will be broadcast on channel *n* with a start time *st* and an end time *en*. The programme will be represented by the sequence of frames

$$p : N \to T$$

A set of instructions must be specified which will achieve the final position of the programme stored on the tape (we shall assume that the tape is initially blank, and the programme is placed at the start of the tape) and the machine in the *Idling* state.

The obvious precondition is that the sequence of operator commands must be completed before the start time *st*. (We are using a particularly simple form of clock with no provision for hours, days, months, years. This can be refined later to produce a more detailed model which would then correspond more completely with a real VCR; this would involve the definition of transformation functions and appropriate types. It is not thought helpful at this stage to do this; as long as it was done precisely, it will pose no real problems as far as the method we are considering goes.)

Theorem
There exists a sequence ρ in seqΣ such that when applied to the machine *M* in the state *Idling*, the final state is *Idling* and the final state of the memory is

$$((n, e, x, y), k)$$

where *k* represents the tape with the programme $f : \{ 1...R \} \to \text{Chan} \times T$ at the start, the current tape position at the end of the programme.

Proof
There are basic assumptions that must be met, and so we assume that *l* is an integer subject to the constraint that *l* exceeds the difference between the initial time value t_0 and the start time *st*. This is to allow us time to carry out the command sequence.

The initial state of the memory is (c, t - 1, (st', en'), ($\perp$, h, 1)), h represents the tape length and is assumed to be sufficient for the accommodation of the programme.

Input	Memory	State
	Chan × T × (T× T) × **Tape**	
	$c, t - 1, (st', en'), (\perp, h, 1)$	*Off*
on	$c, t, (st', en'), (\perp, h, 1)$	*Idling*
ch_up	$c + 1, t + 1, (st', en'), (\perp, h, 1)$	*Idling*
set_prog	$c + 1, t + 2, (st', en'), (\perp, h, 1)$	*Programming*
st	$c + 1, t + 3, (st, en'), (\perp, h, 1)$	*Programming*
en	$c + 1, t + 4, (st, en), (\perp, h, 1)$	*Programming*
set	$c + 1, t + 5, (st, en), (\perp, h, 1)$	*ProgWaiting*
.	.	*ProgWaiting*
.	$c, st, (st, en), (\perp, h, 1)$	*Recording*
.	.	.
.	$c + 1, st+x, (st, en), (f: \{1..r\} \rightarrow \text{Chan} \times \text{T}, h, 1 + r)$	*Recording*
.	.	*Recording*
	$c + 1, en, (st, en), (f: \{1..R\} \rightarrow \text{Chan} \times \text{T}, h, 1 + R)$	*ProgWaiting*
.	.	.
.	.	*ProgWaiting*
unset	$c + 1, en + y, (st, en), (f: \{1..R\} \rightarrow \text{Chan} \times \text{T}, h, 1 + R)$	*Programming*
unset_prog	$c + 1, en + y + 1, (st, en), (f: \{1..R\} \rightarrow \text{Chan} \times \text{T}, h, 1 + R)$	*Idling*

Table 1

The user actions are: on, ch_up, set_prog, *st*, *en*, set, unset, unset_prog

We can also examine what happens if the user fails to complete the instructions before the start time, or if the start-time occurs after the end-time, etc. Appropriate error messages should be produced, but this issue is not addressed further here.

To see what happens with a sequence of user inputs, we can tabulate the results of each action as shown in Table 1. We make various assumptions in this tabulation. Firstly, we have assumed that each series of inputs is entered at a speed of one input for each time point. This is merely to simplify notation. We have not assumed that the tape frames are synchronised with time points. The variables given in the tabulation are assumed to satisfy the following conditions:

$0 < x < st - en$, $0 < y$, $0 < r < R < h$. c is the channel number at the start, assumed to be 1 less than the required channel number n, t the time point at the start and h the length of tape.

Hence, we establish the result.

7.9. EVALUATION AND CONCLUSION

This paper has described a new way of constructing a formal model of a time-dependent system which offers a number of useful opportunities. The model was developed gradually. A simple model of the system was initially constructed and was developed in three stages, each stage providing us with a method of refining an *X*-machine specification, i.e. expanding the state set, expanding the memory and expanding both the state and memory sets (this was achieved by linking two *X*-machines). These methods were used in an intuitive and effective manner to form a complete specification of the system. The paper has also demonstated that the *X*-machine specification can be easily verified against the informal user requirements.

A strong argument that supports the use of *X*-machines is the recent development of a *X*-machine based testing theory([11],[12],[13]), which provides us with a method of constructing test sets from the specification. The technique requires that the *X*-machine specification used satisfies some further conditions. However, this does not restrict the applicability of the method since, by making certain minor modifications, any *X*-machine can be transformed into one that can be fully tested. This design for test approach is an important and valuable aspect of the method which seems quite novel since it encapsulates, in a precise way, the requirements for testability.

The theory of *X*-machines has been developed to a point where it is a useful basis for this work, and this has led us to define an integrated methodology that permits incremental development of a formal specification. This model allows for the discussion of important verification conditions at a high level of abstraction and in an intuitive and productive manner. The existence of a well-founded test set generation method adds dramatically to its usefulness. The success of the model in describing a time-related model in a natural way is a very encouraging sign.

There are still a number of areas where the approach can be improved. Currently, there is no tool support for the process of specifying and analysing *X*-machine specifications. There is no formal proof theory for the formal verification of properties of such specifications, and the notation for the description of the processing functions could be improved. We are currently addressing all of these issues and are developing a number of tools and methods to support the approach. It remains to be seen how easy it will be to produce X-machine based methods and tools which permit the kind of automatic verification of system requirements that the Concurrency Workbench provides.

Future tasks involve exploring more complex examples and case studies and, in particular, safety-related examples. It is also necessary to study refinement and development mechanisms so that there is a firm basis for the construction of a complete design environment that allows us to make the best use of what has been done in the way of outline verification and outline test set generation earlier in the design process at higher levels of abstraction.

REFERENCES

[1] Spivey, J. M., *The Z notation : a reference manual*, Prentice-Hall, 1989.

[2] Jones, C. B. , *Systematic software development using VDM*, Prentice-Hall, 1986.

[3] Goguen, J. A. and Tardo, J. , *An introduction to OBJ: a language for writing and testing software specifications*, In 'Specification of reliable systems', IEEE, 1979.

[4] Milner, R. '*A calculus of communicating systems*', LNCS 92, Springer, 1980.

[5] Hoare, C. A. R. , *Communicating sequential processes*, Prentice-Hall, 1985.

[6] Paulson, L. , *ML for the working programmer*, Cambridge University Press, 1991.

[7] Harel, D. , *Statecharts: a visual formalism for complex systems*, Science of computer programming, 8, 231-274, 1987.

[8] Holcombe, M. Ipate, F., *X-machines with stacks*, Internal report, University of Sheffield, 1994.

[9] Eilenberg, S. *Automata, languages and machines*, Vol. A', Academic Press, 1974.

[10] Holcombe, M., *X-machines as a basis for dynamic system specification*, Software Engineering Journal, 3 (2), 69-76. 1988,

[11] Laycock, G. T., *Theory and practice of specification software testing*, Ph.D. Thesis, University of Sheffield, U.K., 1993.

[12] Ipate, F. Holcombe, M., *X-machine based testing,* submitted.

[13] Holcombe, M Ipate, F., *Formal test set generation*, in 'The Software Process, Testing and Re-use', ed. Tully, C. and Warrilow, D. Unicom, 1995.

[14] Ipate, F. , *X-machine theory*, Internal report, University of Sheffield, 1993.

INDEX

M

O

P

S

T

U

V

X

CPSIA information can be obtained
at www.ICGtesting.com
Printed in the USA
LVHW081819260420
654475LV00017B/945